World Sustainable Energy Days Next 2014

Gerhard Dell • Christiane Egger
Editors

World Sustainable Energy Days Next 2014

Conference Proceedings

 Springer Vieweg

Editors
Gerhard Dell
Christiane Egger

OÖ Energiesparverband
Linz
Austria

ISBN 978-3-658-04354-4   ISBN 978-3-658-04355-1 (eBook)
DOI 10.1007/978-3-658-04355-1

Library of Congress Control Number: 2014953604

Springer Wiesbaden Heidelberg New York Dordrecht London
© Springer Fachmedien Wiesbaden 2015

Printed on acid-free paper

Springer is part of Springer Science+Business Media (www.springer.com)

Foreword from the Editors

The World Sustainable Energy Days are one of Europe's largest annual conferences on energy efficiency and renewable energy and offer a unique combination of events on sustainable energy. The conference, which is organised by the Energy Agency of Upper Austria (OÖ Energiesparverband), is held in Wels in February or March of each year. It attracts more than 700 experts from over 50 countries every year.

A part of the conference is dedicated to young energy researchers. Outstanding young scientists from around the globe present their work in the fields of biomass and energy efficiency. Twenty of the "World Sustainable Energy Days Next 2014" young researchers' presentations are published in this book. The conference was held from 26–27 February 2014.

The papers were selected by a high-level scientific committee for oral presentation. From two main fields—biomass and energy efficiency in buildings—these contributions offer an insight into the research work and the scientific findings and developments of young researchers from all over the world. They also communicate results, trends and opinions that will concern and influence the world's energy experts and policy makers over the next decades.

The conference is organised by the OÖ Energiesparverband, the Energy Agency of Upper Austria. This agency was set up by the regional government to promote energy efficiency, renewable energy sources and innovative energy technologies. Its main target groups are households, public bodies (e.g. municipalities) and businesses. The energy agency is active on the local, regional, national, EU and international levels through numerous projects and programmes. The OÖ Energiesparverband supports the regional government in the development and implementation of regional energy programmes. One of its primary fields of action is comprehensive information and awareness raising activities on sustainable energy production and use. This also includes the organisation of conferences and events, e.g. the World Sustainable Energy Days.

The editors would like to thank all authors of these papers and their academic mentors and supporters. Special thanks go to Professor Reinhold Priewasser from the Institute for Environmental Management in Companies and Regions at the Johannes Kepler University Linz and Dr. Walter Haslinger, Area Manager at Bioenergy 2020+ for their scientific support as well as to the members of the scientific committee. Thank you to Karin Krondorfer for the support in the development of this book and to all of the OÖ Energiesparverband's team who enable the annual reoccurrence of the World Sustainable Energy Days. We also greatly thank the region of Upper Austria for the financial support.

Linz, June 2014   Gerhard Dell and Christiane Egger

Contents

Contributors

Aleksandra Arcipowska Buildings Performance Institute Europe, Brussels, Belgium

Elisa Carlon Bioenergy2020+GmbH, Wieselburg, Austria
Free University of Bozen–Bolzano, Bolzano, Italy

P. Sean Carlson University of Northern British Columbia, Prince George, BC, Canada

Guanyi Chen School of Environmental Science and Technology/State Key Laboratory of Engines, Tianjin University, Tianjin, China

H. D. Doan Tokyo Institute of Technology, Tokyo, Japan

Nicolas Doassans-Carrère Revtech Process Systems, Loriol-sur-Drôme, France

Carol Eastwick Energy and Sustainability Research Division, The University of Nottingham, Nottingham, UK

Karin Fazeni The Energy Institute, Johannes Kepler University, Linz, Austria

P.U. Foscolo Department of Industrial Engineering, University of L'Aquila, L'Aquila, Italy

Prof. Kazuyoshi Fushinobu Tokyo Institute of Technology, Tokyo, Japan

K. Gallucci Department of Industrial Engineering, University of L'Aquila, L'Aquila, Italy

Suyin Gan Department of Chemical and Environmental Engineering, The University of Nottingham Malaysia Campus, Semenyih, Selangor, Malaysia

Laszlo Golicza Bioenergy2020+GmbH, Wieselburg, Austria

Miha Grilc Laboratory of Catalysis and Chemical Reaction Engineering, National Institute of Chemistry, Ljubljana, Slovenia

Bernd Hafner Fachhochschule Burgenland GmbH, Pinkafeld, Austria

Walter Haslinger Bioenergy2020+GmbH, Wieselburg, Austria

Hannes Hebenstreit Fachhochschule Burgenland GmbH, Pinkafeld, Austria

Franz Jetzinger ALPINE ENERGIE Österreich GmbH, Linz, Austria

Vladimir Jovanović Institute of Architecture and Design, Vienna, Austria

Sara Kunkel Buildings Performance Institute Europe (BPIE), Brussels, Belgium

Satu Lantiainen University of Missouri, Columbia, MO, USA

Janez Levec Laboratory of Catalysis and Chemical Reaction Engineering, National Institute of Chemistry, Ljubljana, Slovenia
Faculty of Chemistry and Chemical Technology, University of Ljubljana, Ljubljana, Slovenia

Blaž Likozar Laboratory of Catalysis and Chemical Reaction Engineering, National Institute of Chemistry, Ljubljana, Slovenia
Faculty of Chemistry and Chemical Technology, University of Ljubljana, Ljubljana, Slovenia

Johannes Lindorfer The Energy Institute, Johannes Kepler University, Linz, Austria

Buddhike Neminda Madanayake Energy and Sustainability Research Division, The University of Nottingham, Nottingham, UK

Harald Mattenberger Fachhochschule Burgenland GmbH, Pinkafeld, Austria

Niamh McDonald Global Buildings Performance Network, Paris, France

F. Micheli Department of Industrial Engineering, University of L'Aquila, L'Aquila, Italy

Martin Mitzkat Revtech Process Systems, Loriol-sur-Drôme, France

Sébastien Muller Revtech Process Systems, Loriol-sur-Drôme, France

Denny K.S. Ng Department of Chemical and Environmental Engineering/Centre of Excellence for Green Technologies, The University of Nottingham, Semenyih, Selangor, Malaysia

Hoon Kiat Ng Department of Mechanical, Materials and Manufacturing Engineering, The University of Nottingham Malaysia Campus, Semenyih, Selangor, Malaysia

Rex T.L. Ng Department of Chemical and Environmental Engineering/Centre of Excellence for Green Technologies, The University of Nottingham, Semenyih, Selangor, Malaysia

Viktória Papp University of West-Hungary, Sopron, Hungary

L. Parabello Department of Industrial Engineering, University of L'Aquila, L'Aquila, Italy

Ksenia Petrichenko PhD ResearcherCopenhagen Centre on Energy Efficiency (C2E2) UNEP DTU PartnershipMarmorvej 51,2100 Copenhagen Ø, Denmark

Milan Rashevski Tokyo Institute of Technology, Sofia, Bulgaria

Matteo Rimoldi Bioenergy2020 + GmbH, Wieselburg, Austria
Politecnico di Milano, Milano, Italy

Lasse Rosendahl Department of Energy Technology, Aalborg University, Aalborg, Denmark

L. Rossi Department of Physical and Chemical Sciences, University of L'Aquila, L'Aquila, Italy

Johannes Schmid ALPINE ENERGIE Österreich GmbH, Linz, Austria

Christoph Schmidl Bioenergy2020 + GmbH, Wieselburg, Austria

Markus Schwarz Bioenergy2020 + GmbH, Wieselburg, Austria

Sophie Shnapp Global Buildings Performance Network, Paris, France

Dr. Nianfu Song University of Missouri, Columbia, MO, USA

Wolfgang Stumpf Fachhochschule Burgenland GmbH, Pinkafeld, Austria

Raymond R. Tan Center for Engineering and Sustainable Development Research, De La Salle University, Manila, Philippines

Saqib Sohail Toor Department of Energy Technology, Aalborg University, Aalborg, Denmark

VijayKumar Verma Bioenergy2020+GmbH, Wieselburg, Austria

Theresa Wohlmuth ALPINE ENERGIE Österreich GmbH, Linz, Austria

Zhe Zhu School of Environmental Science and Technology/State Key Laboratory of Engines, Tianjin University, Tianjin, China
Department of Energy Technology, Aalborg University, Aalborg, Denmark

Robust Building Data: A Driver for Policy Development

Sophie Shnapp

If you cannot measure it, you cannot improve it.

Abstract

As the energy performance of buildings is central to any effective strategy designed to mitigate climate change, the building community needs better access to building performance data to improve current policies. This chapter presents the results of research examining the current data quality, data collection best practices and data gaps at the global level based on a desktop survey of ex-post studies and a data quality matrix prepared in collaboration with a group of global experts and modellers.

1.1 The Importance of Building Performance Data

As buildings account for around a third of the global final energy use and 30% of global energy-related carbon emissions, it is clear that this sector has the potential to bestow huge energy savings [1]. For this reason, the Global Buildings Performance Network's (GBPN) mission is to significantly reduce greenhouse gas (GHG) emissions associated with building energy use.

The GBPN work in four priority regions—China, the European Union (EU), India and the United States of America (USA)—together representing around 65% of the global final building energy use in 2005 [1]. The GBPN facilitates this action through regional hubs and partners in the four priority regions. The regional hubs and partners provide the

S. Shnapp (✉)
Global Buildings Performance Network, 51 rue Sainte-Anne, Paris 75002, France
e-mail: sophieshnapp@hotmail.co.uk

© Springer Fachmedien Wiesbaden 2015
G. Dell, C. Egger (eds.), *World sustainable energy days next 2014*,
DOI 10.1007/978-3-658-04355-1_1

most up-to-date knowledge and data on building energy policies to decision makers within their region.

It is estimated that by 2050, if we follow the current policy trends, the energy use from the building sector will increase by around a half of 2005 levels [1]. However, if the current best practices were to become a standard practice, it is possible to reduce global building final energy use by one third of 2005 levels [1].

To build and renovate buildings that are energy efficient and sustainable, participants in the building sector must trust the data used to calculate the energy savings. To gain the confidence of policy makers, builders, architects and all building sector stakeholders, the data must be both available (and storable) and credible (verifiable and transparent). Solid data cases provide known facts that can be used to influence the decision makers; therefore, it is essential that consensus be reached on the basis of credible data collection and its analysis. There is a need for a credible baseline and data series. The baseline is crucial for measuring impact and to oversee if objectives are being achieved.

The quality of data around the world varies considerably; there are large data gaps, weaknesses and inaccessibility that preclude accuracy in modelling. This report presents a unique attempt to assess the quality of data of building types in each of the GBPN's regions. The main aim of the report is to identify the omissions (or "white spots") in the data that prevent modelling and estimation of energy efficiency potentials in buildings. This will assist in the design of measures to improve the quality of data collection and in designing new policies that support a development towards low energy use in buildings. Strategies for overcoming these gaps are provided through advice and reasoned opinions from international experts.

1.2   Methodology

This project has collected information on the quality of data that relates to the energy performance of buildings; the parameters considered for this study were floor area, number of buildings, energy use, heating, cooling, hot water, lighting/appliances, age profile, retrofit rates, urban/rural split, new building energy use, yearly construction, fuel mix, ownership (private/public) and tenure.

All data and information from this report were sourced directly from the GBPN's hubs, partners, regional and global experts and modellers in the four regions and gathered in a data collection matrix. The structure of the matrix consisted of building types down the left-hand column and performance data along the top row. The GBPN's hubs and experts filled in the matrix by scoring each of the parameters with a quality rating between 0 and 5; see Table 1.1. At least two unconnected parties, one global and one regional, filled in each region's data quality matrix.

Table 1.1 Weighting: accuracy descriptions

Weighting	To what degree are the data that you have used accurate?
5	Data source accurate and fully reliable—official verified document or more than one independent source giving similar information
4	Good, trusted data source, i.e. an official document
3	Data generally available, but from mixed sources
2	Partial data—data available not very accurate
1	Weak data—little available data/not accurate
0	No evidence—guess

1.3 Data Quality Findings

The data quality matrices of the four regions give an accurate perspective of how strong or weak the current data quality is. As expected, the quality of data varies significantly between regions although there are some recurring trends. These results gathered in the matrix are presented in a graph below.

The graphed data quality "spider webs" in Figs. 1.1 and 1.2 show the data quality of the four GBPN regions with the different requested parameters for both residential and commercial and public buildings. Generally, the USA has the higher scoring data quality for most of the parameters, followed by the EU and then China and India.

1.3.1 Regional Comparison

At a first glance (Fig. 1.3), it is clear that there are not enough available data in all four regions for accurately modelling building energy performance. It is also clear that the quality of data differs vastly across the regions.

Fig. 1.1 Data quality of four GBPN regions—commercial and public buildings

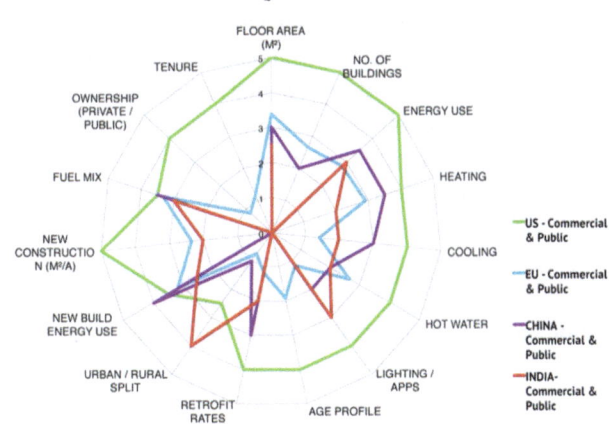

Fig. 1.2 Data quality of four GBPN regions—residential buildings

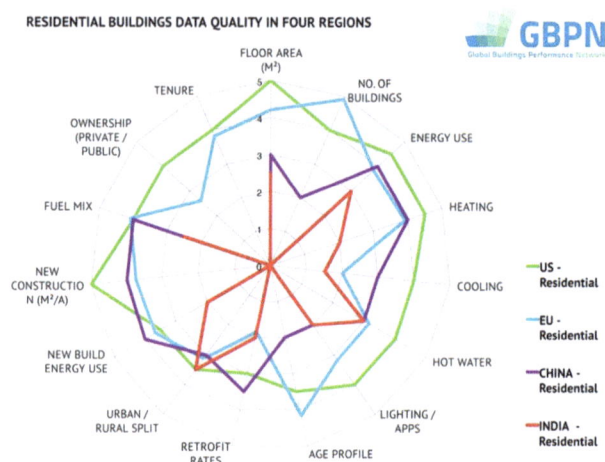

On average, the residential stock scored a rating of 0.5 higher than the commercial building stock (therefore it is 10% more available and accurate than the commercial and residential building stock). When comparing the two graphs, the commercial graph presented more data gaps than the residential.

No region could be considered as having exceptional data as there were significant gaps and weakness found in each region's data set, even after allowing for fields that were not actively investigated by the modellers and experts to be discounted. Figure 1.3 demonstrates the difference of the data qualities in the four regions.

1.3.1.1 Data in China

The residential building data in China scores on average 0.75 times higher than the commercial and public data. China's building performance data averages at around 2.5 in the valuation of the experts, which implies that there are either partial or available data, yet they are not always reliable.

The commercial and public building stock has the lower scoring data compared to the residential data in all the parameters except for seven that are equal. A total of 6 out of the 15 parameters for the residential data are between 3.5 and 5, meaning that on average around a third of the data are deemed as being from a reliable and trustworthy source. The commercial and public building data have four parameters that fall into the "accurate/reliable" weighting category, this means a quarter of the data were weighted as being accurate. The rest of the data are not found to be accurate or even available.

1.3.1.2 Data in the EU

Unlike the USA, the EU does not have official data on the building sector as a whole region and the quality of data varies significantly between the different states; therefore, the EU results are taken from an average of six of the member states—two countries that

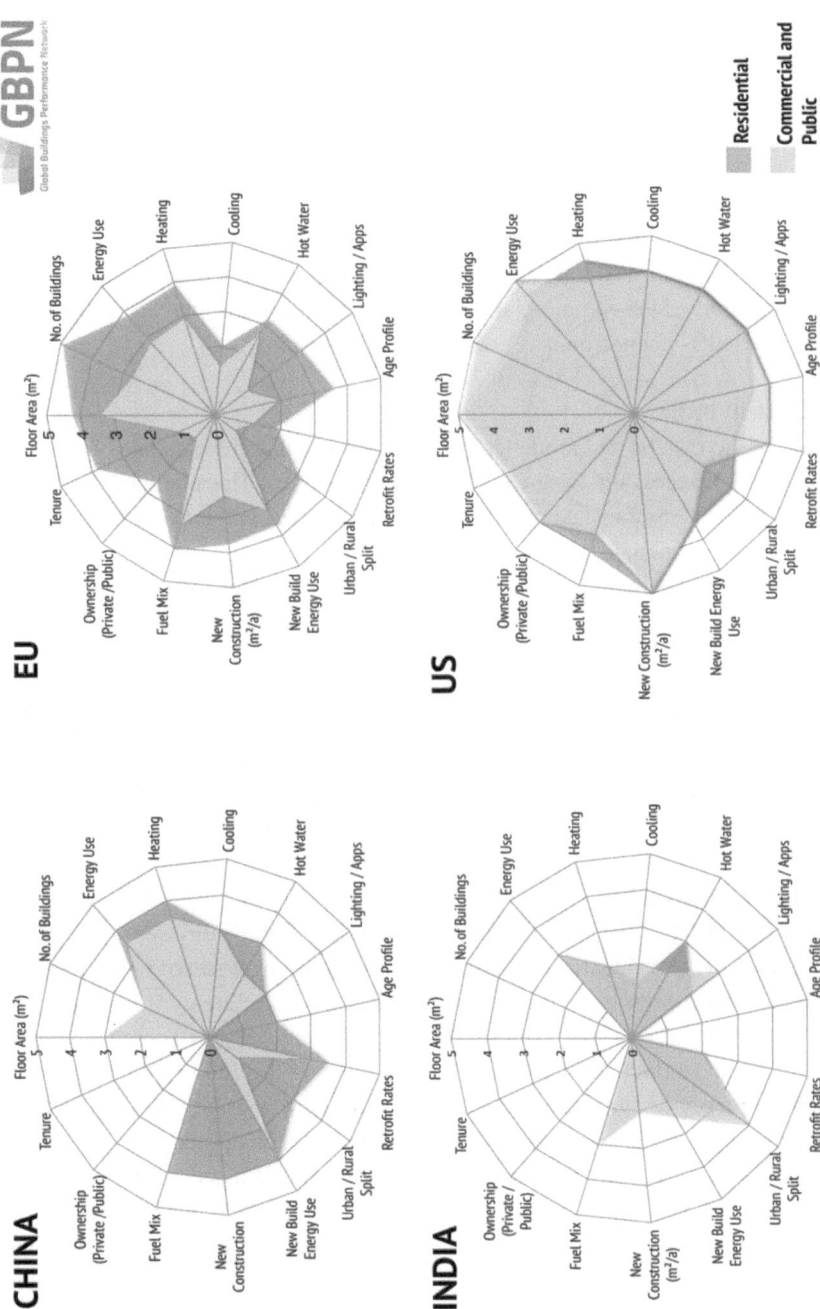

Fig. 1.3 Residential versus commercial data quality in the four GBPN priority regions

are below the EU "average", two average EU countries and two countries above the EU average; the countries used for this study were Austria, Germany, Poland, Spain, Sweden and the UK. The findings of the analysis found that the quality for both residential and commercial and public buildings in the EU is lower than the USA.

The biggest difference between the quality of residential and commercial buildings is found in the EU; the average score of the residential sector was higher than the average commercial sector by just under two, giving a 40% difference between the two sectors. In general, around two thirds of the residential data are deemed as being accurate (these parameters were found to be in the top two weighting categories). The commercial and public sector has the lower-scoring data compared to the residential data in all the parameters. For almost half of the commercial and public data, there are sources available, yet they are not deemed as being from a reliable or accurate source.

1.3.1.3   Data in India

The response of the experts and modellers in India showed that the data used for modelling are frequently inaccessible for the modellers of the survey. This could be due to data being difficult to locate or translate from the original language, and it might be a result of the very diverse and regional structure of India. Although it was possible to estimate the existing residential and commercial building data to give an understanding of how energy use is split by end use, the experts weighted the accuracy of data on average as 1.7 out of 5 and therefore they seem to be unreliable.

The commercial and public stock has stronger data compared to the residential data in all the parameters, except two that are equal and one that is higher. Only one of the parameters for both commercial and residential were scored as having accurate data; 4 out of the 15 parameters from the commercial building sector have a score of 2.1–3.4, meaning that the majority of the data were deemed as being unavailable or inaccurate.

1.3.1.4   Data in the USA

On a scale of 0–5, the US commercial and public building data quality score on average was 0.06 higher than the residential data. The USA has the strongest set of building energy data among the GBPN priority regions; this is supported by the Energy Information Administration's (EIA) national-level data surveys on the characteristics and energy use of commercial and residential buildings (EIA 2013). However, the 2007 CBECS data were withheld due to survey design issues, and the 2011 survey was briefly postponed due to federal funding cuts, meaning the latest available data are over a decade old.

Out of the four regions, the residential and commercial data sets in the USA were the closest together with a difference of 0.06. The commercial and public stock has the strongest data compared to the residential data in all the parameters except three. Eleven out of the 15 parameters (for both commercial and public and residential buildings) have a weighting of 3.6–5, meaning that, in general, the US building data are available and often the source is trusted—the source is taken from an official or accurate place (in this case, most of the data are taken from the official energy statistics retrieved from the US Government—the EIA).

1.3.2   Understanding the Differences in Data Quality

Collecting data in multiple regions with different cultures, languages and political contexts is complex. Regions vary in their laws, standards, definitions and values connecting to data collection. The main reason for the data varying widely across the four regions is due to the different political approaches to data collection.

The findings of this research show that that in some regions, data collection methodologies are more advanced than others. Some regions have taken the first steps towards assessing data by completing specific surveys and setting up collection frameworks; for other regions, collecting data is more complex. The demand for data is still low in developing countries; this makes the collection process more difficult. Some regions collect data at a national level and some at a more local level, the differences in levels create further difficulty in having consistency.

1.4   Main Recommendations for Better Data Collection Practices

The GBPN has a group of international experts working in the field of building sector energy efficiency. Thirty of GBPN's experts in the field of building energy data contributed to the survey on data sources, availability and quality.

The survey provided a valuable opportunity to gather expert opinion on how to improve data quality and collection around the world, and more specifically, in the GBPN's four key regions. As well as allowing for a more comprehensive understanding of how we can collaboratively improve data, there were a number of recurring recommendations that became obvious when analysing the advice from the experts from all regions.

There is overwhelming emphasis on the need for a comprehensive data collection framework to ensure the consistency of data. Many experts also strongly advised that data collection definitions and guidelines should be harmonised and clarified. Repeatedly, the main themes included:

- The need for a comprehensive framework;
- The need for data collection definitions and guideline;
- A collaborative effort to share data and begin the provision of open source data;
- The need for a collection and analysis tool (comprehensive database);
- The need to make data collection mandatory;
- The need for incentives and funding; and
- A dissemination of data collection best practices and case studies.

1.5   What Conclusions Can Be Drawn from This Report?

What is clear is that regionally there are large differences in data quality and that large data gaps exist, making it difficult to analyse the current state of play in each region. Substantial efforts need to be made to fill these data gaps and inaccuracies. Although data are not deemed as being accessible, it does not necessarily mean they are not available or cannot be found, for instance by local actors in this region, but it demonstrated a need for an improvement of data access for these modellers.

We must continuously advance our collection techniques to harmonise and improve access to secure building energy data, alike. There is a need to share available data more broadly. Initially, it is essential to prioritise our most pressing needs regarding the most crucial data required by modellers and policy makers.

Since no one group of experts can do everything, there is a need to work together in order to make a difference. A collaborative approach towards ensuring transparency of data must be adopted so that data collection, monitoring, reporting and evaluation leave no gaps and produce accurate and reliable data. Data collection should be harmonised so that national and regional data collection systems relating to the energy performance of buildings are consistent.

The GBPN calls for a collaborative effort in harmonising definitions and measurement templates concerning the energy performance of buildings. Furthermore, the GBPN recommends the establishment of a database that includes all the building energy performance data and the diverse requirements of the building stakeholders.

References

1. ÜRGE-VORSATZ, D., et al. (2012) *Best Practice Policies for Low Energy and Carbon Buildings. A Scenario Analysis*. Research report prepared by the Center for Climate Change and Sustainable Policy (3CSEP) for the Global Buildings Performance Network. Paris.

Aleksandra Arcipowska

Abstract

As the energy- and climate-related discussion evolves and becomes more complex, policy makers need more and better data to design and evaluate policies and programmes.
The significance of data input determines the quality of decisions taken. Data are also
the key element that allows making comparisons and establishing the monitoring systems to track the progress and impact of various policies. Currently, there is no official and centralized database on the European buildings stock. The buildings data are
collected by different institutions (i.e. statistics offices, energy agencies, consultancy
companies, research organisation, others) mainly on the member state level; its quality,
availability and completeness varies significantly between the different countries.

One of the key challenges for the European buildings data collection is the issue of
data harmonisation. Currently, even though various data are available from different
sources (both official and unofficial) their cross-country comparison is very hard to
conduct. Moreover, there is a need to better use of existing tools for data collection (i.e.
central engineering, procurement and construction (EPC) and renovation registers),
and also use of new, smart tools for data collection (especially for non-residential sector) and sharing.

A. Arcipowska (✉)
Buildings Performance Institute Europe, Rue de la Science 23, Brussels, Belgium
e-mail: aleksandra.arcipowska@bpie.eu

© Springer Fachmedien Wiesbaden 2015
G. Dell, C. Egger (eds.), *World sustainable energy days next 2014,*
DOI 10.1007/978-3-658-04355-1_2

2.1 The Need for Buildings Data

People spend more than 90% of their lifetime in buildings; these are our homes, schools, hospitals and governments, and are places to work and spend leisure time. We spend not only our time in buildings but also our money on buildings. It is no surprise that we (building owners, tenants) would like to know more about buildings and their impacts.

One of the most interesting building-related aspects is energy consumption. It is directly linked to comfort and health; it influences the size of our wallets and has an impact on the environment. The existing building stock corresponds to roughly 40% of the total primary energy use and more than 36% of the total energy-related greenhouse gas emissions. The building sector has the highest potential for energy efficiency improvements (in comparison to other sectors of economy) and occupies a central place in the political debate on a low-carbon future.

As the energy- and climate-related discussion evolves and becomes more complex, policy makers need more and better data to design and evaluate policies and programmes. The significance of data input determines the quality of decisions taken. Data are also the key elements that allow making comparisons and establishing the monitoring systems to track the progress and impact of various policies.

The need for credible buildings data is not only limited to policy makers, regulators and standards developers. As the construction sector is one of the key drivers of the European economy, buildings data have a great value for the market. They are the reference point in the business strategies for various parties including real estate agents, manufacturers, design professionals, etc.

2.2 The Current Status of the European Buildings Data

In 2013, the Global Buildings Performance Network published a comparative study on the buildings data robustness in the EU, China, India and the USA [1]. The authors of the report investigated and compared the quality and availability of the buildings data for the residential and non-residential sector. The results indicate that there is a room for improvement for European buildings data.

Currently, there is no official and centralized database on the European buildings stock. The buildings data are collected by different institutions (i.e. statistics offices, energy agencies, consultancy companies, research organisation, others) mainly on the member state (MS) level; its quality, availability and completeness varies significantly between the different countries. There are only a few building indicators (e.g. construction rates, final energy consumption in household and services sector) that are collected and harmonised by the official European Statists Office, the Eurostat [2, 3].

In 2010, the Buildings Performance Institute Europe (BPIE) decided to launch an ambitious and a long-term research project aiming at the harmonisation of European buildings data. The first step was to learn about existing data sources, data providers and most importantly data needs. In 2010/2011, an extensive survey was conducted to collect in-depth statists and policy information for 29 European countries (EU 27+CH, NO). Results have been published in the report entitled Europe's buildings under the microscope [2] and at the BPIE data hub, an online portal on the European buildings data (www.buildingsdata.com). In the second part of 2013, BPIE decided to extend the scope and update the available information on the portal. In November (for the first anniversary of the data hub), the first survey 2013 has been launched. The BPIE work on data collection (Survey 2011, 2013) allows drawing conclusions on the current status of buildings data across Europe (see Fig. 2.1).

The most reliable data are available for the buildings stock statistics for the residential sector (i.e. floor area, total number of buildings/dwellings/units, owner occupancy profile, etc.). This information is based mainly on the population and housing census data (conducted every 10 years in every MS on the basis of the Regulation (EC) No. 763/2008). Official buildings stock statistics for the non-residential buildings (especially for commercial buildings) are limited and available only in some of the EU countries.

An important part of the building's statistics is energy performance data. The key document that sets a framework for data collection on energy consumption on buildings is Regulation (EC) No. 1099/2008 on energy statistics. Even the European Commission is trying to stimulate the member states to collect detailed data on the final energy consumption in households (by using a direct survey: interviews, e-mail/web questionnaire) or to develop a model for estimating energy consumption in households, but the availability of disaggregated information (by end use, building type, age bands) is still very limited. In 2013, Eurostat published a manual for statistics on energy consumption in household that introduces the methodologies and best practices of data collection across Europe [4]. It show that even more than 75 % of the data available are based on the results of surveys and modelling exercises, there is a growing importance of administrative data (such as cadastre register, energy performance certificate registers) and in situ measurements (see Fig. 2.2).

An important source of information on energy-efficient trends in sector of households, tertiary, industry and transport is the ODYSSEE project, supported by the Intelligent Energy Europe Programme of the European Commission. The ODYSSEE database was been established in the 1990s, and among many others, it presents disaggregated information on energy consumption in the household sector by end use and by building type, as well as takes into account climatic correction of data. ODYSSEE is based on the network of national data providers (EU 28+Norway) who work together on data harmonisation and quality check. Free access to the full database is restricted to the EU policy makers and non-profit organisation (from the begging of 2014).

There are also other EU-funded projects that provide information on buildings' energy performance, such as ENTRANZE, TABULA or EPISCOPE. For example, in the EPISCOPE project (a continuation of the TABULA project), disaggregate information on energy consumption in residential buildings has been gathered (i.e. by building type and

Building stock data

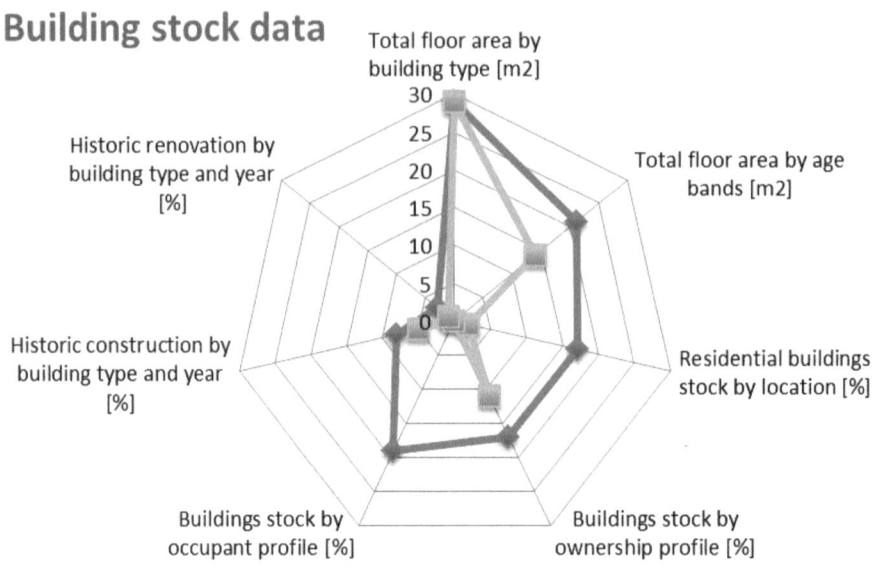

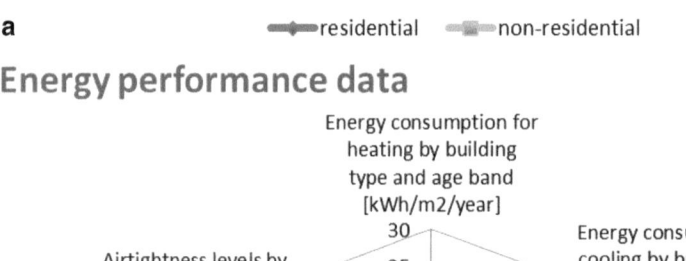

a   ➤residential   ➤non-residential

Energy performance data

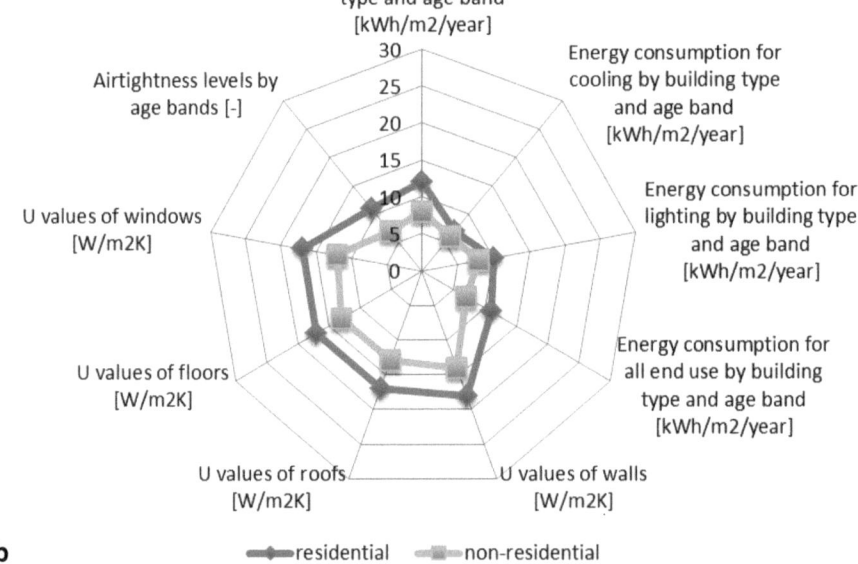

b   ➤residential   ➤non-residential

Fig. 2.1 The European buildings data availability for residential and non-residential buildings

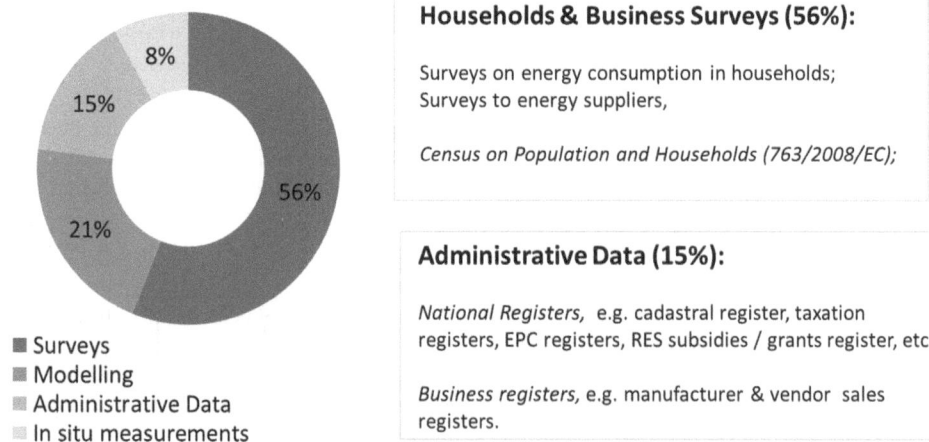

Households & Business Surveys (56%):

Surveys on energy consumption in households;
Surveys to energy suppliers,

Census on Population and Households (763/2008/EC);

Administrative Data (15%):

National Registers, e.g. cadastral register, taxation
registers, EPC registers, RES subsidies / grants register, etc.

Business registers, e.g. manufacturer & vendor sales
registers.

■ Surveys
■ Modelling
▨ Administrative Data
▨ In situ measurements

Fig. 2.2 Energy performance data acquisition in Europe [4]

by age band). Moreover, average U-values for the building components are estimated (i.e. by age band and by building type).

Besides the statistical information on the European buildings stock, there is a need to collect policy-related information. In this field, there is a high dynamic of information change which requires suitable methods of the data collection and verification. In recent years, the growing importance to disseminate policy-related information on buildings was the Concerned Action group that supports the technical, legal and administrative framework for the Energy Performance of Buildings Directive (EPBD). One of the activities of the group is to provide an overview on current status of the EPBD implementation across Europe [5, 6].

Comprehensive information on the building-related policy information (i.e. building codes, framework for the energy performance certificate system, economic and market-support instruments) is available at the BPIE data hub. At the portal, information can be searched in customised manner and compared across the countries. Another source is the Mesures d'Utilisation Rationnelle de l'Energie (MURE) database that presents a wide range of the energy efficiency policy measures for the household and service sectors.

2.3   The Challenges and Opportunities of the Buildings Data Collection in Europe

Over the past year, as a consequence of the political discussion on the EPBD recast and Energy Efficiency Directives (EED), increased attention has been given to buildings data. Comparing the results of the BPIE survey of 2010 and 2013, it can be concluded that both availability and quality of buildings data have increased. Nevertheless, on the European buildings data map, there are still major gaps that need to be addressed in the coming years.

Table 2.1 Examples of useful floor area definition in various EU MSs [2]

Hungary	*Overall useful area surrounded by plastered or tiled walls, where the inner height exceeds 1.90 m. In multi-dwelling buildings, beside total area of flats, it comprises the area of public premises as well*
Lithuania	The sum of floor areas of all heated premises of a building, including floor areas of heated basements, stairwells, shared premises and other heated premises, as well as floor areas of premises, all walls of which are shared with heated premises
Poland	Useful floor area of a dwelling includes area of all rooms, kitchen, bathroom, etc
Portugal	The sum of the areas, measured from the inner perimeter of the walls of all compartments of a building unit of a building (based on the building drawing), including lobbies, corridors, internal sanitary facilities, storage and other interior compartments function similar cabinets and walls

2.3.1 Harmonisation of the European Buildings Data

One of the very important challenges for the European buildings data collection is the issue of data harmonisation. Currently, even though various data are available from different sources (both official and unofficial), their cross-country comparison is very hard to conduct. This is because the methodologies of data collection vary between countries. Member states have different buildings categories, age bands, but most importantly, different definitions are taken into account (e.g. definition of a useful floor area, see Table 2.1).

2.3.2 Better Use of Existing Tools for Data Collection (Central EPC and Renovation Registers)

The energy performance certification and inspection programmes, developed in the process of implementing the EPBD in 2002, have the potential to yield comprehensive data on the energy performance of buildings. Even though the recast of the EPBD in 2010 does not stipulate registers, central EPC registers are useful tools to implement the mandatory independent controls. In May 2014, 24 out of 28 MS had implemented central registers on national or regional ECP. These central databases vary in regards to scope, format and the accessibility rights.

The Portuguese case study [7] shows that the national databases of buildings energy certification registers can be extremely useful in obtaining statistically relevant insights on the energy performance of the existing buildings stock. They can also allow analysing the trends and verify to what extent the historical evolutions are justified by natural evolution or by the effects of regulations.

In the long-term perspective, it seems to be one of the most cost-effective solutions for the data collection. The key challenge for European buildings data is to address the opportunity of systematic data collection via central EPC registers, and also to make data publicly available. The challenges and benefits of making EPC data publicly available were analysed among others by the decision makers in the UK [8].

An additional potential source of information on buildings is the long-term renovation strategies that MSs need to prepare by 30 April 2014 (on the basis Art 4 of the EED directive). The analysis of the current status quo of existing building stock and its performance is a starting point for this kind of strategic document.

2.3.3   Use of New Tools for Data Collection (Especially for Non-Residential Sector)

Taking into account the gaps in the European buildings data for the non-residential buildings, an innovative method of data collection should be designed. One good example that could be followed is the US initiative is called the Green Button. The project is based on cooperation between industry and the White House. Within its scope, energy utilities provide their (residential and commercial) customers with smart tools that allow access to the energy usage information. Users can monitor their energy consumption in the building, and at the same time, data are gathered in the central database. The critical success factors of the project are based on the well-constructed solution for the private data protection, good cooperation with energy utilities and the simplicity of the tool. It is enough to "click on the green button" and download the free software to benefit from knowing more on the real behaviour of the building.

2.4   How to Make European Buildings Data Useful

The process of collecting credible and comprehensive information on buildings is probably of utmost importance to respond to data needs. Nevertheless, it is not the only condition that must be satisfied for making buildings data useful (for policy makers, building owners, building professionals, etc.).

One of the most important criterion to increase data usefulness is to assure free access to data (i.e. licence free). Nowadays, the poor data availability is not necessarily resulting from data gaps, but rather from limited access to data (i.e. paid or restricted access for specific parties).

Another condition (once data are publicly available) that need to be satisfied is to make buildings data useful and reusable. This can be assuring with the standardised format of data representation; it is easy to imagine that e.g. "tables" are more user-friendly when given in "xls" (or "csv") file rather than a pdf format. Suitable formats are important both for data sharing and for data acquisition and notably machine readable formats for ease of automation.

★	Information is available on the Web (any format) under an open license
★ ★	Information is available as structured data (e.g. Excel instead of an image scan of a table)
★ ★ ★	Non-proprietary formats are used (e.g. CSV instead of Excel)
★ ★ ★ ★	URI identification is used so that people can point at individual data
★ ★ ★ ★ ★	Data is linked to other data to provide context

Fig. 2.3 Five starts model of LOD concept [9]

All issues mentioned above are supported with the Linked Open Data (LOD) concept [9], which, in my opinion, is the desirable architecture for the future of the European buildings data collection. The LOD principle is not only about assuring an open access to data but most importantly to allow an easy exchange, reuse and linked of data (see Fig. 2.3).

References

1. SHNAPP, S., LAUSTSEN, J. (2013) *A Robust Building Data: A driver for policy Development – Technical report*, Global Building Performance Network.
2. ECONOMIDOU, M., et al. (2011) *Europe's Buildings under the Microscope. A country-by-country review of the energy performance of buildings*, The Buildings Performance Institute Europe.
3. BROIN, E. (2007) *Energy Demands of European Buildings: A Mapping of Available Data, Indicators and Models, Thesis for the Degree of Master of Science in Industrial Ecology*, Chalmers University of Technology.
4. EUROSTAT (2013) *Manual for statistics on energy consumption in households*, Publications Office of the European Union.
5. CA EPBD (2010), *Implementing the Energy Performance of Buildings Directive* (EPBD). Featuring Country Reports 2009, Publications Office of the European Union, ISBN 978-972-8646-27-1.
6. CA EPBD (2013) *Implementing the Energy Performance of Buildings Directive* (EPBD). Featuring Country Reports 2012, ADENE, ISBN 978-972-8646-28-8.
7. MAGAKHAES, S., LEAL, V. (2014), *Characterization of thermal performance and nominal heating gap other residential building stock using the EPBD-derived databases: The case of Portugal mainland*, Energy and Buildings 70 (2014) 167–179.
8. DEPARTMENT FOR COMMUNITIES AND LOCAL GOVERNMENT (2012); *Making energy performance certificate and related data publicly available*. An impact assessment, DCLG, ISBN 9781409834410.
9. SEMANTIC WEB COMPANY (2010); *Linked Open Data: The Essentials. A Quick Start Guide for Decision Makers*.

Dynamic Building Energy Codes: Learning from International Best Practices

Niamh McDonald

Abstract

If greenhouse gas emissions from the building sector are to be significantly reduced, all new buildings must be built to zero-energy standards. Mandatory dynamic building energy codes integrated into long-term strategies are necessary if these standards are to be met. This chapter presents a set of specially developed criteria that define the state of the art in building energy codes. These criteria form the basis of an interactive tool that facilitates the comparative analysis of 25 best practice building energy codes. This tool supports policy makers to develop more ambitious policies by outlining how current best practices may be improved to in light of the criteria developed.

3.1 Introduction

Energy use in buildings is responsible for more than 30 % of CO_2 emission and has a significant role to play in climate-change mitigation. This can be achieved through a number of strategies, with energy efficiency providing a low cost solution [5]. A study completed by the Central European University on behalf of the Global Buildings Performance Network (GBPN) has demonstrated how far a transformative change of the building sector can bring us in terms of emissions reductions, if building energy efficiency levels are significantly increased [5]. In order to achieve this "deep" scenario, today's best practice/state-of-the-art buildings must become the standard in less than 10 years from now. For new and existing buildings, this means that all buildings should develop towards net zero

N. McDonald (✉)
Global Buildings Performance Network, 51 rue Sainte Anne, 75002 Paris, France
e-mail: nm@gbpn.org, niamhmcdonald@gmail.com

© Springer Fachmedien Wiesbaden 2015
G. Dell, C. Egger (eds.), *World sustainable energy days next 2014,*
DOI 10.1007/978-3-658-04355-1_3

energy. Energy efficiency codes and supporting policies play a critical role in ensuring reduced energy consumption for the life of the building [1, 2].

Despite the existence of a vast number of policies worldwide, the world is unlikely to reduce its greenhouse gas (GHG) emissions to keep the global temperature increase below the 2 °C target as compared to the preindustrial level [4]. It has been suggested that in order to achieve greater results, energy-efficiency policy making must be more dynamic in terms of a continuous closed-loop process that "involves and balances policy design, implementation and evaluation" [3]. A report by Lawrence Berkley National Laboratory [3] commissioned by the GBPN indicates that an upscale of current practices is not enough to bring us to the "deep" scenario and that there is a need to increase actions significantly. It is suggested that the key to altering current trends is to prescribe mandatory, dynamic and ambitious building codes and supporting policy packages that are incorporated into long-term strategies. It is clear from the report that such requirements are essential in ensuring the support of the market in implementing best practices in energy-efficient buildings.

3.2   Methodology

3.2.1   Overview

The GBPN works to implement of the "deep" scenario whereby today's state-of-the-art buildings are the norm in less than 10 years from now. In order to support this mission, in 2013, GBPN developed a project with the aim of defining "state of the art" in building energy code development and analysing current best practices in light of the definition. "State of the art" was defined using 15 key criteria, and 25 current examples of best practice building codes were analysed in light of these criteria. The analyses are represented in an interactive online tool, which allows users to analyse the codes based on individual criterion or using multiple criteria. The tool aims to facilitate the sharing of best practices between jurisdictions by analysing the various elements of the different codes.

3.2.2   Key Themes, Criteria and Sub-Questions

With the support of 64 international building energy efficiency experts, GBPN began the process of defining a "state-of-the-art" building energy code by identifying five key themes. The five themes identified as critical to a state-of-the-art building code include: Holistic Approach, Dynamic Process, Implementation, Technical Requirements and Overall Performance. Based on the five themes identified, a detailed set of criteria were developed. These criteria not only form the basis of a "state-of-the-art" building code but also facilitate the rigorous assessment of codes and supporting policies. An iterative process of feedback and refinement was followed and consensus was eventually reached on 15 criteria. The criteria are defined in Table 3.1.

Table 3.1 State-of-the-art themes and related criteria

Theme 1: *A holistic approach to buildings*	Theme 2: *A dynamic process*	Theme 3: *Proper implementation*	Theme 4: *Technical requirements*	Theme 5: *Overall performance*
1. Performance-based approach	4. Zero-energy target	7. Good enforcement	10. Building shell	13. Overall performance/on site
2. Performance to include all energy types or uses	5. Revision cycles	8. Certification	11. Technical systems	14. Overall performance/primary energy
3. Energy efficiency and renewable energy	6. Levels beyond minimum	9. Policy packages	12. Renewable energy systems	15. Overall performance/GHG emissions

GHG greenhouse gas

Once consensus was reached on both key themes and criteria, a set of sub-questions for assessing each of these 15 criteria were developed. These questions were designed to investigate in further detail all aspects of the individual criterion, which in turn supported the scoring of the codes in light of the criteria.

3.2.3 Scoring Principles

In order to accurately compare best practice building energy codes, it was necessary to develop a scoring system that reflected the true nature of existing policies. As outlined above, the sub-questions for each criterion were used as the basis for scoring the codes. Each criterion was assigned a maximum score of 10 points with the 10 points distributed amongst the sub-criteria depending on the importance of each question. A score of 10 was awarded to examples of absolute best practice or perfect development in this field.

The scores awarded to each code are illustrated on the GBPN website in the Policy Tool for New Buildings and can be reviewed by all users. Users are encouraged to "play" with the tool, selecting and deselecting criteria/elements that are of interest and comparing the 25 codes selected by GBPN based on those criteria. The main aim of the tool is to learn from the scoring under each criterion rather than determining which code scores best overall. Fig. 3.1 provides an illustration of the tool.

3.2.4 Selection of Best Practice Building Energy Codes

One of the aims of the policy tool is to promote examples of dynamic and ambitious building energy-efficiency codes. In light of this aim, 25 energy-efficiency codes from across the GBPN regions (China, Europe, India and USA) and some examples on codes from jurisdictions outside of the GBPN regions were selected. All the codes represented best practice in their region and in their level of development. Codes were selected following a literature review of current best practice. The GBPN regional hubs, IMT, BPIE and part-

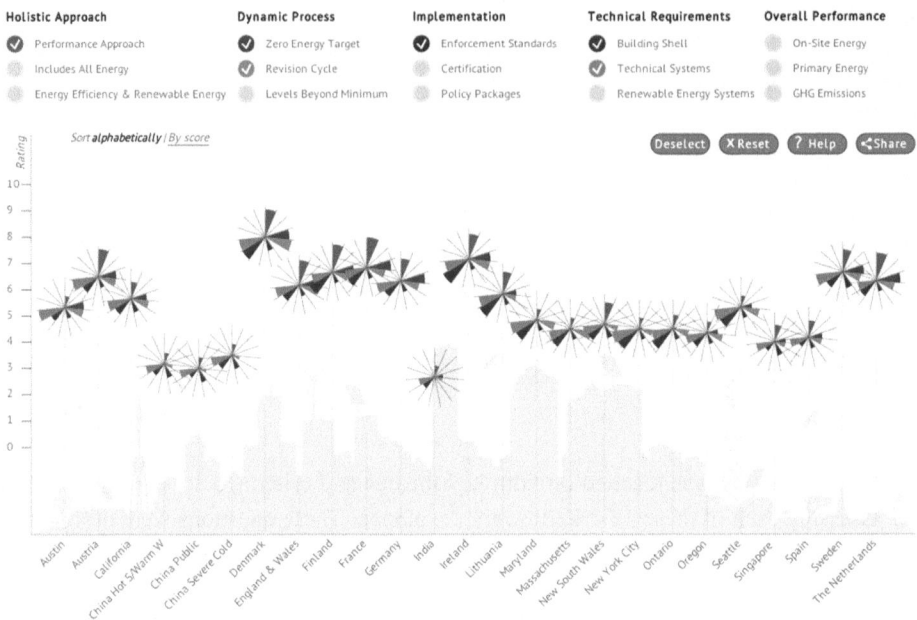

Fig. 3.1 Image of Global Buildings Performance Network (*GBPN*) policy tool for new buildings

ners, Shakti and GBPN China, provided considerable support in the selection of codes. Codes included can be seen in Fig. 3.1.

3.3   Results

Results from the tool are discussed in the following section.

3.3.1   Performance Approach

Significant energy improvements can only be achieved by focusing on the overall performance of the building and the balance between the individual elements. The codes that scored best under this theme shifted from an elemental approach to code design to a mandatory overall performance-based approach.

France scored full marks under these criteria due to its extremely progressive performance-based code. The code requires buildings to demonstrate compliance with the maximum allowed primary energy consumption, the "Cepmax" coefficient. The code was also awarded full marks due to the requirements for mandatory renewable energy requirements, computer simulations, air-tightness testing for residential buildings and bioclimatic design considerations, all of which are integral to a holistic approach.

The Netherlands, Finland, Denmark and Austria also scored very well due to similarly progressive approaches, but other European codes did not score that well due to the overall performance calculation being a model or equivalent building calculation. This was also the case for some of the US codes. The best-performing codes from other regions were those of New South Wales and California. Despite not having mandatory overall calculations, these codes take GHG emissions or peak loads into account and generally encourage bioclimatic design strategies.

3.3.2 Dynamic Process

While many codes lacked realistic and achievable zero-energy targets, other codes lacked binding roadmaps on which these targets should be based. Issues also existed surrounding the definition of zero energy and the setting of aspirational targets for revisions towards zero energy.

The best-performing jurisdictions under this theme were those that have set a nearly zero-energy building (nZEB) target, complete with roadmap and frequent revision cycles in order to reach those targets. They included the Netherlands, Denmark, France and Germany, all of which are required under the EU legislation to ensure that, by 2020, all new buildings are built to nearly zero-energy standard. The highest score across this theme was awarded to Denmark due to its long history of frequent update cycles and the inclusion of clearly defined future targets in addition to well-established complementary policies such as energy performance certification. For example, under the 2008 Danish Building Regulations, two low-energy classes were defined that have/will become the minimum requirements by 2010 and 2015. The 2011 regulations (BR10) introduced a further low-energy class that will be the requirement by 2020. Such long-term targets provide the market sufficient time to prepare for the coming changes.

Codes from Massachusetts, Oregon and Sinapore scored well with regard to levels beyond minimum. Under the theme of dynamic process, Austin, California and Oregon scored well as they demonstrated dynamic code development processes that are moving towards zero energy. However, room forfurther strengthening of this process still exists.

3.3.3 Enforcement

The analysis highlighted significant issues with enforcement of the building code in all jurisdictions. The highest score awarded under this category was 5/10 and that was awarded to Sweden due to its post-occupancy energy-verification compliance option. Where this compliance option is selected by the municipal building board, an interim permit of use is granted for the first 2 years of occupancy. Where non-compliance is found to exist, the developer must desist from using the building until the issues are corrected. The compliance checking is closely related to the energy performance certificate, which is also granted based on the metered energy consumption of the building.

Denmark, California, New South Wales and the Chinese standards all scored 4/10, with all other codes scoring 3/10. The low scores are generally due to a lack of compliance statistics and energy verification requirements. None of the codes was found to have robust and independent compliance monitoring and all codes can significantly improve in this area.

3.3.4   Overall Performance

Fully performance-based codes or performance track compliance options have been adopted or offered in relatively few of the selected best practice codes. As a result, it has not been possible to score all of the codes under the theme "Overall Performance." This has emerged as a problem in particular in India, China and the USA. The lack of overall performance values for many of the codes emphasises the need for a more holistic approach to code development that focuses on the overall performance of the building including primary energy, GHG emissions and all end uses.

In order to score all of the codes as fairly and as accurately as possible, it is intended to develop a series of performance values based on a number of model reference buildings.

3.4   Conclusions

The interactive policy tool is an innovative tool that clearly outlines the mechanisms behind dynamic building codes and policies. It facilitates the comparison and analyses of best practice energy codes using a clearly defined set of criteria. This analysis and comparison provides those involved in policy development with a clearer picture of the key elements necessary for the design of best practice codes and of how best practice codes are implemented in practice. By providing detailed insight into 25 examples of current best practice codes, this tool supports policy makers in the development of more dynamic and ambitious building codes that can move the building stock towards zero energy.

As is clear from the findings of this research, there is still much work to be done in order to move the building stock towards zero energy in less than 10 years from now. The "Policy Tool for New Buildings" can be used to facilitate this progress by assisting those involved in policy development to develop best practice solutions based on lessons learnt from different countries and regions. The tool demonstrates that there are multiple approaches that can be adopted to achieve zero energy, but all should address the five key themes outlined in this chapter. Using these themes as a foundation provides policy makers with the flexibility to search for the practical solutions that are best suited to their individual jurisdiction.

References

1. HANSEN KJÆRBYE, V., LARSEN, A., TOGEBY, M. (2011) Do changes in regulatory requirements for energy efficiency in single-family houses result in the expected energy savings? [Presented at ECEEE Summer Study] Preque Isle de Giens, June 2011.
2. JACOBSEN, G. D., & KOTCHEN, M. J. (2013) Are building codes effective at saving energy? Evidence from residential billing data in Florida. *Review of Economics and Statistics*, *95*(1), 34–49.
3. LEVINE, M. et al. (2012) Building Energy Efficiency: Best Practice Policies and Policy Packages. *Lawrence Berkeley National Laboratory*. California: Global Buildings Performance Network.
4. UNEP (2012) *The Emissions Gap Report 2012—A UNEP Synthesis Report*. Nairobi: United Nations Environment Programme (UNEP).
5. ÜRGE–VORSATZ, D. et al. (2012) *Best Practice Policies for Low Energy and Carbon Buildings. A Scenario space Analysis*. Budapestz: Center for Climate Change and Sustainable Policy (3CSEP) for the GBPN.

Towards 2020: Zero-Energy Building for Residential and Non-Residential Buildings

4

Hannes Hebenstreit, Bernd Hafner, Wolfgang Stumpf and
Harald Mattenberger

Abstract

This chapter assesses the options for both non-residential and residential buildings to reach a zero-energy building (ZEB) standard. Besides the already widely practiced consistent minimization of the energy demand for heating and hot water supply of buildings, further progress can be made by minimizing the electric energy demand. Besides highly efficient lighting, the calculations done showed a significant savings potential by using energy-efficient electrical devices.

4.1 Introduction

The topic of nearly zero-energy buildings (NZEBs) has received increasing attention in recent years, even becoming part of the energy policy in several countries. In the recast of the EU Directive on Energy Performance of Buildings [1], it is specified that by the end of 2020, all new buildings shall be "NZEBs". However, despite the emphasis of the EU targets [2], these remain generic in most cases and are not yet standardized. The aforementioned EU Directive provides the following definition:

"Nearly zero-energy building" means a building that has a very high energy performance, as determined in accordance with Annex I. The nearly-zero or very low amount of energy required should be covered to a very significant extent by energy from renewable sources, including energy from renewable sources produced on-site or nearby. [1]

H. Hebenstreit (✉) · B. Hafner · W. Stumpf · H. Mattenberger
Fachhochschule Burgenland GmbH, Steinamangerstrasse 21, 7423 Pinkafeld, Austria
e-mail: bernd.hafner@fh-burgenland.at

© Springer Fachmedien Wiesbaden 2015
G. Dell, C. Egger (eds.), *World sustainable energy days next 2014*,
DOI 10.1007/978-3-658-04355-1_4

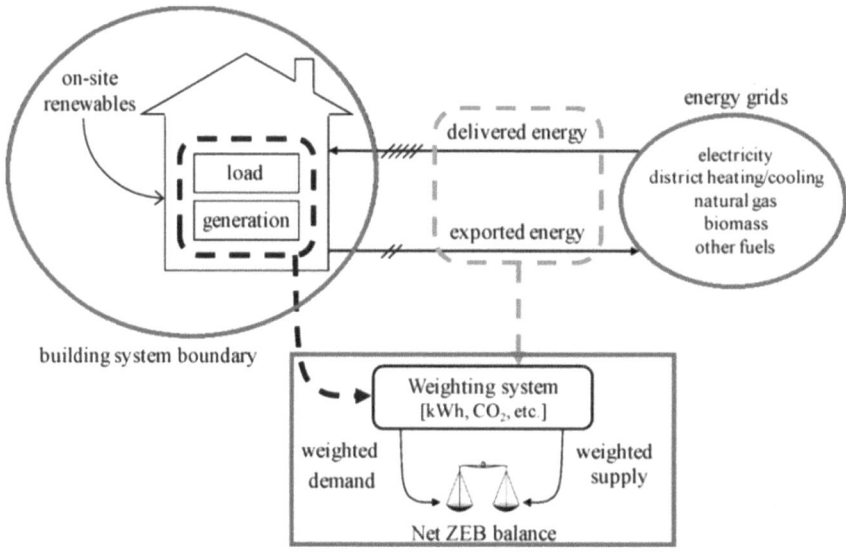

Fig. 4.1. Important terminology in a NZEB balance [5]

The definition of zero-energy goals affects the choices building designers make to achieve these goals and whether they can claim success. The term zero-energy building (ZEB) often is used commercially without a clear understanding, and countries are enacting policies and national targets based on the concept without a clear and unanimous definition. However, the common denominator for the different NZEB definitions is the balance between weighted demand and supply. This relationship is shown in Fig. 4.2.

An NZEB still remains connected to the public grid and imports needed or exports surplus electrical energy depending on current conditions of supply and demand within its boundary.

4.2 Methodology

In this chapter, an analysis to reach a ZEB balance for both a non-residential building (a school) and a residential building (a single family home) based on a time span of 1 year is presented and discussed.

In a first step, these two analyses focus on the comparison of the final and the primary non-renewable energy demand of a selection of calculated cases. The data for the conversion factors are taken from the OIB guideline [3] and represent the Austrian legal requirements. Included are the data regarding single contributions by different energy demands for the initial state (equal to "worst case") and a series of possible improvements (towards "best case"). The improvements will be elaborated in the results chapter. In a second step, the primary non-renewable energy demand as well as the energy costs and the CO_2-emis-

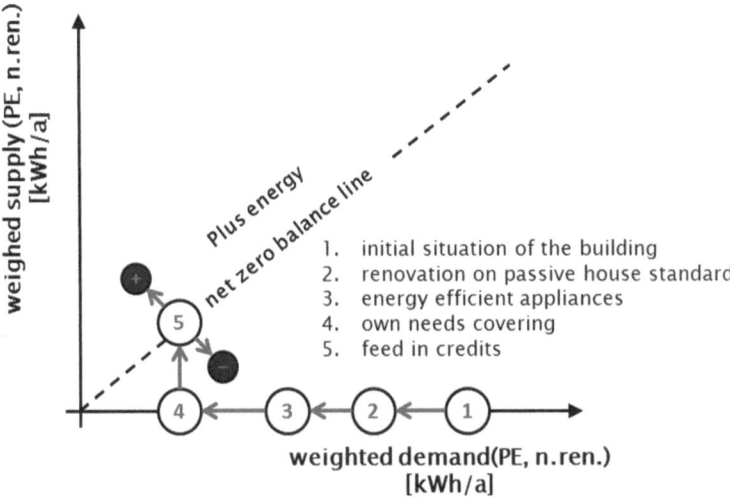

1. initial situation of the building
2. renovation on passive house standard
3. energy efficient appliances
4. own needs covering
5. feed in credits

Fig. 4.2 NZEB balance method [6]

sion equivalents are displayed in a graph as described in Fig. 4.2. Energy costs are based on data from proPellets Austria [4].

For illustration, the most important terminology like the building system boundary of the NZEB, the energy grids and the weighting system are shown in Fig. 4.1.

Figure 4.2 shows the used energy balance methodology of this chapter to present the results of the analysed examples. The graph shows the weighted demand on the X-axis and the weighted supplies on the Y-axis. The dashed diagonal line denotes ZEB conditions. All calculations are based on cumulative monthly final energy data generated with the simulation tool "ETU Gebäudeprofi Duo" [7] and the individually programmed tool "RE-ACT support tool for calculation of ZEB". The latter allows—amongst other features—a detailed calculation of the energy demand of individual electrical devices. The simulation tool "Polysun" [8] is used to calculate the photovoltaic (PV) system necessary to achieve a zero-energy demand within the building system boundary.

4.3 Results and Discussion

4.3.1 Non-residential Building: School with Three Wings

The object analysed is the School of Ceramics in Stoob (Austria), which consists of three wings, all built in the late 1960s: the two-storey school building, the single-storey work-shop and the two-storey boarding school. Natural gas is used for heating purposes; electric power is taken from the public grid. The heat demand of the building stock is about 86 kWh/m²$_{GFA}$ a. For illumination, solely light bulbs are in use. The rooms are ventilated manually using windows. To evaluate the current requirement of electrical energy in the

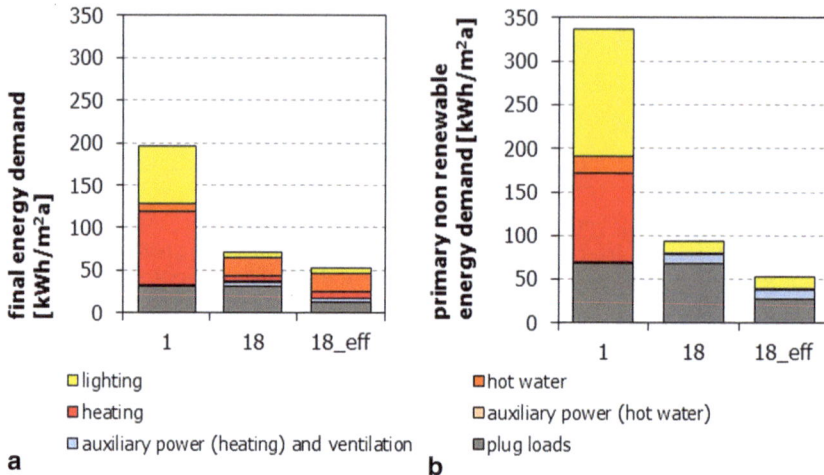

Fig. 4.3 a Final and **b** primary non-renewable energy demand of the school buildings for three selected cases (as described in the text)

buildings, inefficient electrical equipment was assumed using representative consumption data [9].

In a first step, the energy demand of the building stock just described is valued by the building simulation tool and the results are represented in case 1. In a next step, 17 alternative cases are calculated and analysed (cases 2–18). Parameters to be varied include the insulation of the building envelope and windows, types of heating systems, ventilation, lighting and control possibilities as well as further electrical devices (plug loads).

In Fig. 4.3a, b, the final and primary non-renewable energy demands for three different cases are shown. These cases represent the worst- and best-case scenarios regarding the parameter variations chosen. In the existing school building (case 1), the energy demand for heating takes the largest proportion (44%), followed by lighting (35%) and plug loads (16%). In order to reach the low overall energy demand of case 18, several steps are taken:

- The insulation of the building envelope is optimized to reach passive house standard.
- A ventilation system including heat recovery is installed in order to reduce the ventilation heat losses to a minimum and to ensure a healthy air quality.
- The remaining heat demand is covered by a new biomass boiler.
- The outdated light bulbs are replaced by a new light-emitting diode (LED) lighting system.
- The decentralised hot water system is substituted by a central hot water heating system (exhibiting more distribution losses), but is driven by biomass instead of electricity as energy carrier.

Case 18 features the same building envelope and building equipment as case 18_eff. The only difference is the use of less efficient versus more efficient devices (plug loads) with

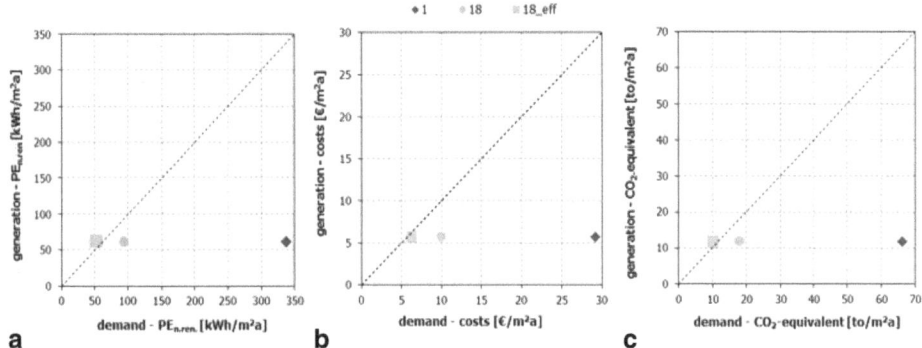

Fig. 4.4 Building balance results regarding **a** primary non-renewable energy demand, **b** costs and **c** emissions expressed as CO_2 equivalents

respect to their electrical energy consumption. Through the consistent use of these devices, the energy demand in that category can be reduced by 60%.

The analysis of the final energy demand (in Fig. 4.3a) shows that the total demand (based on case 1) can be reduced by 63% (in case 18) without using efficient devices (PC, printer, etc.). If highly efficient devices are employed in addition, an overall reduction of 73% can be accomplished (in case 18_eff).

In Fig. 4.3b, the primary non-renewable energy demand of the considered cases is depicted. Because of the high conversion factors for electricity and natural gas, the primary non-renewable energy demand is about 60% higher than the final energy demand for case 1. Although the final energy demand for hot water rises (see Fig. 4.3a), the primary non-renewable energy demand (see Fig. 4.3b) can be reduced by using biomass instead of electricity from the grid. This is caused by the fact that the conversion factor from final to primary non-renewable energy demand for biomass is much lower compared to the factor for imported electricity from non-renewable sources. Looking at the primary non-renewable energy demand of case 18 and 18_eff, the demand for heating and hot water almost disappears. Here, the biggest consumers become the plug loads that are influenced by the user of the building. Through the use of efficient devices, the primary non-renewable energy demand of case 18 can be additionally reduced by 44%, as depicted in case 18_eff.

All in all, for the primary non-renewable energy demand, savings of 72% (for case 18) and 84% (for case 18_eff) can be achieved in comparison with case 1. To reach the goal of a zero-energy, zero-emission or zero-costs building, it is necessary to generate energy from on-site renewables. Therefore, the whole rooftops of the buildings are used to install a PV system with a total power outlet of 180 kWp, which represents the maximum installation power possible. Looking at Fig. 4.4a, the balance between demand and generation can only be reached if efficient devices are used. If also case 18 should reach the diagonal line, the PV system would have to be enlarged. On the one hand, this would lead to higher investment costs and, on the other hand, more space for the PV system would be necessary.

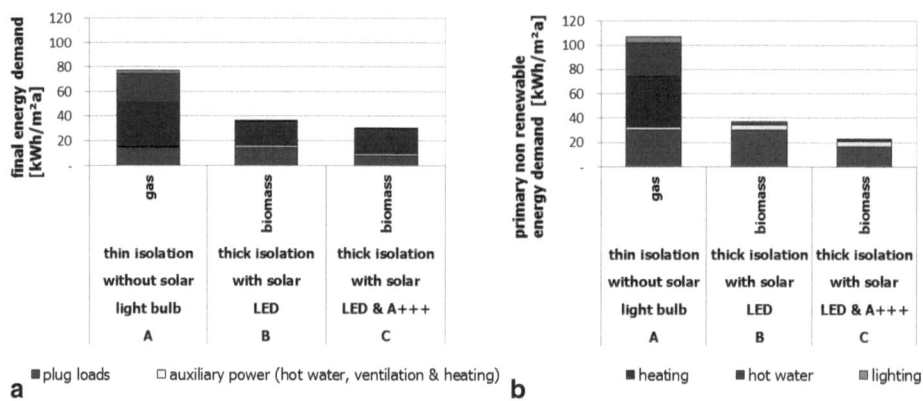

Fig. 4.5 **a** Final and **b** primary non-renewable energy demand of the analysed single-family house for three selected cases (as described in the text)

4.3.2 Residential Building: Two-Storey Single-Family House

The analysed building is a typical two-storey single-family house with a gross floor area of 173 m². For the investigation, two different heating systems (natural gas boiler and biomass boiler) with and without a solar thermal system (with solar fraction of 20% for hot water and heating demand) are analysed. Furthermore, the envelope quality is varied from a standard isolation (with an opaque U-value from 0.2 W/m²K) to a passive house standard isolation (with an opaque U-value from 0.1 W/m²K). Every system of these analyses presented herein includes a ventilation system with a heat recovery rate of 75%.

In Fig. 4.5a, b, the final and primary non-renewable energy demands for three different cases are shown. These cases represent the best- and worst-case scenarios regarding the parameter variations chosen. The worst case (A) of the analyses shows the system (natural gas) with a standard building envelope quality. The final energy demand for this case shows the highest demand for heating (47%), followed by the energy demand for hot water (31%), plug loads (18%) and lighting (3%). For this, worst-case plug loads with an energy-efficiency labelling standard of class "A" and a lighting system with light bulbs (60 W power) are employed.

In a next step (case B), the envelope wall quality and the windows are changed to passive house standard, which reduces the heat losses through the building envelope. Moreover, the energy source for heating is changed from natural gas to biomass and, in addition, a solar thermal system for heating and hot water production is installed, which reduces the entire final energy demand by 36%. The installed solar system leads to a hot water supply that requires only 2% of the total energy demand in this case. Also, the lighting system is changed from light bulbs to a modern LED lighting system, which reduces the energy demand for lighting by approximately 85%.

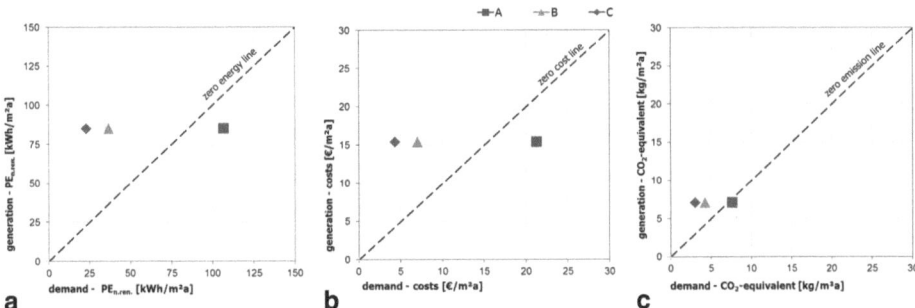

Fig. 4.6 Building balance results regarding **a** primary non-renewable energy demand, **b** costs and **c** emissions expressed as CO_2 equivalents

In the final step (case C), the plug loads are changed from the energy-efficiency labelling class "A" to "A+++" devices, which reduces the electrical energy demand for plug loads by 45 %.

The analysis of the final energy demand in Fig. 4.5a shows that the total demand of energy can be reduced by 61 % from worst-case scenario (A) to best-case scenario (C). In Fig. 4.5b, the primary non-renewable energy demands of the three cases are shown. Due to the high conversion factors for electricity and natural gas, the energy demand in case "A" rises by 28 % from 77 kWh/m²a (final energy) to 107 kWh/m² a (primary non-renewable energy) a year.

The energy demand for heating and hot water supply in both cases B and C is so minute that these demands are almost invisible in the graph (less than 1 % of the total energy demand). Hence, the electrical energy demand gets more and more important. In case B, the total energy saving potential, compared to case A, is around 66 % and in case C around 78 %.

In case C, the necessary on-site energy production, in order to reach a zero-energy, zero-emission or zero-costs building, is assured by a rooftop PV system with an installation power of 5 kWp, which is a standard PV array for a single-family house. To reach this value, approximately 40 m² of PV array is necessary to generate about 5600 kWh/a at the location Pinkafeld (Austria).

Figure 4.6a,b, c shows the results of the analysed examples represented by the explained methodology at the beginning. The results show clearly that only systems A and B reach the goal of a zero-energy, zero-emission and zero-costs building. The case A does not reach the balance of demand and generation. If also case A should reach the zero energy line, the PV system would have to be enlarged by 1.5 kWp. On the one hand, this would lead to higher investment costs and, on the other hand, more space for the PV system would be necessary.

4.4   Conclusions and Outlook

The analyses in this work show important key factors to reach a zero-energy, zero- carbon and zero-costs building for residential and non-residential buildings.

The first and most important step to reach the ZEB goal is the reduction of the energy demand to a minimum. Both sample buildings have shown that the zero-energy goals can only be reached by a consistent reduction of the energy demand for heating as well as electricity. Thus, the key to achieve the zero-energy building goal is an effective insulation of the building envelope, a ventilation system with a heat recovery system, a renewable energy source like biomass for the supply of the remaining heating energy demand as well as highly energy efficient lighting (LEDs) and devices with the energy labelling A+++. Nevertheless, the human factor that governs the time of use of plug loads cannot be ignored.

Acknowledgment The research was performed in the international project "REACT—Renewable Energy and Efficiency Action" funded by the European Regional Development Fund (ERDF), creating the future—programme for cross-border cooperation between Slovakia and Austria, 2007–2013.

EUROPEAN UNION
European Regional
Development Fund

References

1. EPBD 2010/31/EU (2010) Directive 2010/31/EU of the European Parliament and of the council of 19 May 2010 on the energy performance of buildings. *Official Journal of the European Union.* [Online] L 153. p. 13–35. Available from: http://eur-lex.europa.eu/LexUriServ/LexUriServ.do?uri=OJ:L:2010:153:0013:0035:EN:PDF [Accessed: 01 January 2014].
2. Eu Kommission (2011) *Energieeffizienzplan 2011: Mitteilung der Kommission and das europäische Parlament, den Rat, den europäischen Wirtschafts- und Sozialausschuss und den Ausschuss der Regionen.* [Online] KOM (2011) 109 p.1–19. Available from: http://eur-lex.europa.eu/LexUriServ/LexUriServ.do?uri=COM:2011:0109:FIN:DE:PDF. [Accessed: 18 January 2014].
3. Oib Richtlinie 6 (2011) *Energieeinsparung und Wärmeschutz.* Österreichisches Institut für Bautechnik. Wien.
4. Propellets Austria (2014) Netzwerk zur Förderung der Verbreitung von Pelletsheizungen. Available from: http://www.propellets.at/de/pelletpreise/preisvergleiche [Accessed: 18 March 2014].
5. Sartori, I., Napolitano, A. and Voss, K. (2011) *Net zero energy buildings: A consistent definition framework.* Elsevier—Energy and Building. 48. 220–230.
6. Voss, K. and Mussal, E. (2011) *Nullenergiegebäude—Internationale Projekte zum klimaneutralen Wohnen und Arbeiten.* Detail Green Books. München.
7. Etu Gebäudeprofi Duo (2013) *Gebäudeprofi Duo Energieausweissoftware.* ETU GmbH. Wels.
8. Polysun (2013) *Polysun Designer Simulation Software, Version 6.5.* Vela Solaris AG. Winterthur.
9. Nipkow, J., Gasser, S., Bush Wissen, E. (2007) *Geräte im Haushalt. Informationen und Broschüren.* [Online] Available from: http://www.aew.ch/wissen/interessierte/wissenscenter/strommessgeraete.html [Accessed: 01 January 2014].

Toolbox to Design Housing Refurbishment

Vladimir Jovanović

Abstract

The purpose of this chapter is to assess the effectiveness of individual refurbishment measures for common detached houses in Serbia and establish a design matrix of solutions. Dynamic simulations are performed for three typical houses in two cities. The results are arranged in a form of a toolbox where the measures are classified by their effectiveness. The toolbox could be used as a supportive instrument throughout design and serve as a pre-step in the house renovation in Serbia.

5.1 Introduction

Designing a refurbishment is an important issue for the European building industry, because recent studies indicated that about 80% of the existing buildings will still be present by the year 2050 [1]. According to the Serbian Energy Agency, households had a share of 35% in total energy consumption in Serbia in 2012 [2]. The First Serbian National Energy Efficiency Action Plan (NEEAP) indicates that the average energy consumption in the residential sector is approximated to 220 kWh/m²a [3]. Actions taken in early design phase strongly influence the effectiveness of the entire refurbishment. To create applicable tools that assist designers to develop better retrofits, more research efforts in this field are necessary.

Number of studies has analysed the refurbishment potential in Serbia including authors as Bojic [4]. These researches have been mostly analysed in specific case studies. Nevertheless, Konstantinou developed a methodology for a systematic approach to the

V. Jovanović (✉)
Institute of Architecture and Design, TU Wien, Gußhausstr. 28-30, 1040 Vienna, Austria
e-mail: jovanovic.vlad@yahoo.com

© Springer Fachmedien Wiesbaden 2015
G. Dell, C. Egger (eds.), *World sustainable energy days next 2014,*
DOI 10.1007/978-3-658-04355-1_5

refurbishment of multistorey buildings in European countries (i.e. Netherlands and Germany) [5]. Her work contained a toolbox of specific measures and their effectiveness, thus the measures could be reapplied on many multistorey buildings. So far, no one appears to have applied current knowledge of systematic retrofit to the field of house refurbishment in Serbia. Based on the current findings, we will apply the toolbox model for passive retrofit measures regarding single-family houses. Considering houses from the 1970s and 1980s, we set research questions:

- To evaluate the effectiveness of individual energy conservation measures.
- To deliver the outcomes in a form of a toolbox, which could be used to support designing a refurbishment.

The rest of this chapter is organized as follows: Sect. 5.2 presents the methodology; Sect. 5.3 contains the results; Sect. 5.4 shows the applicability of the toolbox; pros and cons are discussed in Sect. 5.5; and concluding marks are given in Sect. 5.6.

5.2   Research Procedure

Our study investigated the effectiveness of single retrofit measures for detached houses in Serbia, in Southeast Europe. Our methodology was developed in accordance to the Konstantinou's method [5]. We used case study research, experimental and simulation research tactic.

In an attempt to clarify the effect of the single refurbishment measures, we applied them on three typical Serbian houses positioned in two different cities. The analysed locations were the capital, Belgrade, with 2520 heating degree days (HDD); and the city in the southeastern part of the country, Niš, with 2613 HDD [6]. This approach enabled the estimation of the average effectiveness of the measures and improved preciseness of the outcomes. The single measures were systematically arranged in a form of a "toolbox" (Table 5.1).

Because heating demands have the highest share in Serbia's energy consumption, we used heating demands as an energy efficiency indicator [3]. To estimate heating demands of the houses and evaluate the effectiveness of single measures, simulation software Euro-Waebed was used. This tool took into account the presence of people, heat gains from the equipment as well as the air infiltration. For dynamic simulations, the indoor temperature during the heating period was set to 20 °C. The software provided an appropriate accuracy of output data and it has been already used in works published by previous authors [7].

For the reason that 43 % of the Serbian housing stock was emerged in the period 1971–1990, we focused on this group of buildings [8]. Houses from this period are characterized by a similar concrete structure, brick-block walls, cement plaster and a lack of a thermal insulation. We selected three house models from a catalogue of typical housing designs from the 1970s and 1980s in Serbia and other ex-Yugoslavian Republics. We determined typical houses "P+1−6", "P+1−2/1" and "HP+1−116" to be the simulation models (Fig. 5.1) [9].

Table 5.1 Toolbox of energy conservation measures

External walls	Floor	Roof	Windows	Thermal bridges	Air supply
No insulation	No basement	Roof ceiling—not insulated	Single glazing	Linear thermal bridges (e.g. balconies)	Window ventilation
Little/ outdated insulation	Floor on the ground—not insulated	Roof ceiling—insulated	Double uncoated	Geometrical thermal bridges (e.g. junctions)	Ventilation with heat recovery
External insulation ETICS 10 cm	Floor on the ground—insulated	Pitched roof—not insulated	Upgrade existing windows	Repeating thermal bridges (e.g. joists)	
ETICS advanced (EPS 20 cm)	Basement ceiling—not insulated	Pitched roof—insulated	Replacement: 2x glazing	Thermal bridges— partly insulated	
Internal insulation	Basement ceiling—insulated below slab	Green roof	Replacement: 3x glazing	Thermal bridge-free construction	
Ventilated facade			Shading		

EPS expanded polystyrene

Two-storey house
"P+1-6" in Belgrade

Two-storey house
"P+1-2/1" in Nis

Two-storey house with basement
"HP+1-116" in Belgrade

Fig. 5.1 Three houses extracted from the catalogue [9]

Based on the described models, the measures were applied separately by changing only one building element in the basic model. This was done with a purpose to determine individual contribution of each single measure (Table 5.1). For the external wall insulation, ceilings and roofs, we applied the expanded polystyrene (EPS); for the inner insulation we used breathable insulation panels; for the floors on the ground extruded polystyrene foam (XPS). Replacing the windows considered double-glazed (4×12.4 mm) and triple-glazed windows ($4 \times 8.4 \times 8.4$ mm), both with the krypton gas filling. For the ventilation, we used heat recovery system with the effectiveness of 95 %. According to these settings, we performed dynamic simulations that provided output data for further evaluation.

5.3 Results

Analysed houses differed by their energy demands. Our investigation showed that the heating demands for the basic houses varied from 107 kWh/m²a to 153 kWh/m²a. This was due to the diverse number of floors, building form and volume. Although the applied measures showed similar effectiveness, the "P+1−6" model showed slightly better results due to the less thermal bridging (i.e. concrete balconies) in the construction. The upgrade of the building features considered the improving their thermal properties, a reduction of the overall heat-transfer coefficient (U-value), renewing openings with high quality components and installing mechanical ventilation with heat recovery. The energy saving potential of the measures is stated in the Table 5.2. Retrofitting the external walls demonstrated a considerable saving potential. The savings only by this measure were up to 40%. The floor upgrade showed 11% savings comparable to 8% by basement ceiling insulation, though we should remark that the insulating the basement slab is simpler to perform. The roof ceiling insulation showed up to 15% of the savings. Interestingly, the equal reduction of the energy demands was achieved with double-glazed and triple-glazed windows. Regarding our models that are explained in Sect. 5.2, the double glazing was an optimal measure for the windows replacement. The ventilation system itself reduced the initial energy consumption for 15% on average.

A comparison of the saving potential of the individual measures is shown in the Fig. 5.2. Notice that every measure on the external walls demonstrated greater potential than the upgrade of any other building component. This supports that the walls are the basic element to be treated in the refurbishment. Less effective than wall renovation were heat recovery ventilation, roof ceiling insulation, windows replacement and floor insulation, respectively.

5.4 Case Study: Refurbishing Two-Storey House

The showcase presented the application of toolbox for the refurbishment. The case was previously described "HP+1−116" house (Sect. 5.2). The toolbox was used to diagnose the initial condition, plus to develop two retrofits by choosing set of measures for the specific case. In the initial state, the house was in concrete structure with brick-block walls without any insulation. The U-value of the non-insulated external wall was 0.80 W/m² K. The windows had low performance and the envelope was not airtight. In the first refurbishment scenario, we intended to achieve 50% of the savings and we applied following measures: an upgrade of 10-cm EPS insulation on the walls and the roof ceiling plus the installation of double-glazed windows. In the second scenario, we aimed to reach the passive house standard, thus reduce the energy demands by "factor 10". For that purpose, the walls and the roof ceiling were refurbished with 20-cm EPS insulation, the basement slab and the roof skin with 10-cm EPS insulation, the airtightness was improved, double-

Table 5.2 Energy savings by single measures

Building feature	Retrofit measure	U-value[W/m²K]	P+1−6130 kWh/m²a (%)	P+1−2/1 153 kWh/m²a (%)	HP+1−116107 kWh/m²a (%)	Average saving of the measure (%)
External walls	External insulation 10 cm	0.26	34	32	32	33
	External insulation 20 cm	0.16	40	38	38	39
	Internal insulation 10 cm	0.263	32	30	27	30
	Ventilated facade (10 cm insulated)	0.26	34	32	32	33
Floor	Ground-floor on the ground 10 cm	0.33	11	11	–	11
	Basement-floor on the ground 10 cm	0.33	–	–	<1	1
	Basement ceiling 10 cm	0.26	–	–	8	8
Roof	Roof ceiling 10 cm	0.26	13	10	11	11
	Roof ceiling 20 cm	0.16	15	12	13	13
	Roof skin (10–14 cm)	0.30	7	6	7	7
	Green roof	0.55	5	5	5	5
Windows	Double glazing (4 × 12.4 mm, Kr)	1.10	7	12	15	11
	Triple glazing (4 × 8.4 × 8.4 mm, Kr)	0.70	7	12	15	11
Air supply	Ventilation with HR 95%	–	16	12	17	15

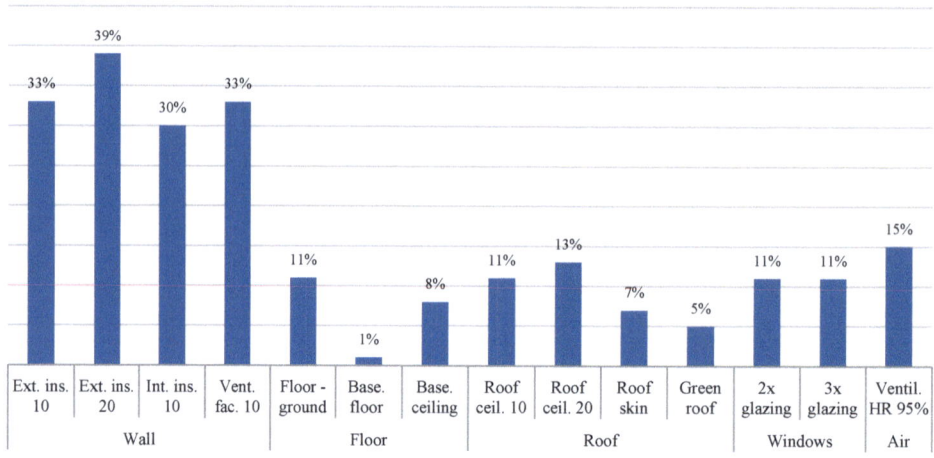

Fig. 5.2 Energy saving potential of single measures

glazed windows with shutters, plus the 95 % efficient heat-recovery ventilation system were installed. The scenarios are marked in the toolbox shown in the Table 5.3.

The initial consumption of the house was 107 kWh/m²a. The consumption in the first scenario was reduced to 45 kWh/m²a. Notice the reduction of the heating demands for 58 and 90 % after the refurbishments (Fig. 5.3). In the R2 scenario, factor 10 refurbishment was achieved, where the consumption of 11 kWh/m²a met the national passive house standard. This test showed the significance of the systematic approach to the refurbishment design.

5.5 Discussion

We identified the effectiveness of the single energy conservation measures and used the case study to manifest how the systematic knowledge can facilitate planners to design suitable refurbishments.

Our research procedure related to the cited method [5], thus proposed valuated design approach for the detached houses in the southeast European climate. A greater number of samples could lead to a higher generalization of our observation. The toolbox could provide first-step information about the retrofit potential, though each house is a particular and requires a specific treatment.

Table 5.3 Toolbox for diagnosis the current state, R1 and R2 refurbishment

WALL	FLOOR	ROOF	WINDOW	THERM. BRIDGE	AIR	WALL	FLOOR	ROOF	WINDOW	THERM. BRIDGE	AIR	WALL	FLOOR	ROOF	WINDOW	THERM. BRIDGE	AIR
No insulation	No basement	Roof ceiling- no ins.	Single glazing	Linear th. bridges	Window ventilation	No insulation	No basement	Roof ceiling- no ins.	Single glazing	Linear th. bridges	Window ventilation	No insulation	No basement	Roof ceiling- no ins.	Single glazing	Linear th. bridges	Window ventilation
Outdated insulation	Ground-floor- no insulation	Roof ceiling- insulated	Double uncoated	Geometric bridges	Ventilation with heat recovery	Outdated insulation	Ground-floor- no insulation	Roof ceiling- insulated	Double uncoated	Geometric. th. bridges	Ventilation with heat recovery	Outdated insulation	Ground-floor- no insulation	Roof ceiling- insulated	Double uncoated	Geometric bridges	Ventilation with heat recovery
ETICS standard 10	Ground-floor – insulated	Pitched roof – no ins.	Upgrade existing windows	Repeating bridges		ETICS standard 10	Ground-floor - insulated	Pitched roof - no ins.	Upgrade existing windows	Repeating th. bridges		ETICS standard 10	Ground-floor - insulated	Pitched roof - no ins.	Upgrade existing windows	Repeating t. bridges	
ETICS advanced 20	Basement ceiling – no ins.	Pitched roof – insulated	Replace: 2x glazing	T. bridges - partly insulated		ETICS advanced 20	Basement ceiling - no ins.	Pitched roof - insulated	Replace: 2x glazing	T. bridges - partly insulated		ETICS advanced 20	Basement ceiling - no ins.	Pitched roof - insulated	Replace: 2x glazing	T. bridges - partly insulated	
Internal insulation	Basement ceiling – insulated	Green roof	Replace: 3x glazing	Th.-bridge-free		Internal insulation	Basement ceiling - insulated	Green roof	Replace: 3x glazing	Th.-bridge-free		Internal insulation	Basement ceiling - insulated	Green roof	Replace: 3x glazing	Th.-bridge-free	
Ventil. Facade			Shading			Ventil. facade			Shading			Ventil. facade			Shading		

Fig. 5.3 Heating demands
of the house (kWh/m²a)

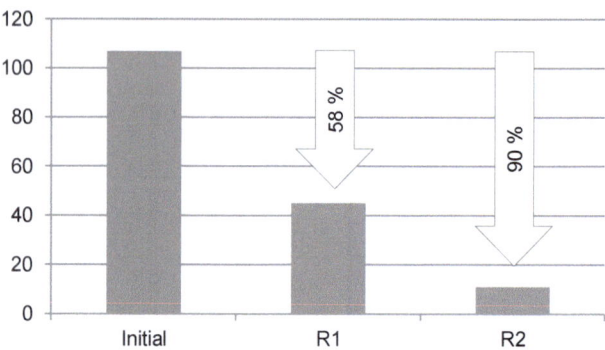

5.6 Conclusion

The purpose of this work was to analyse individual effectiveness and establish a toolbox of house renovation measures. The intention was to demonstrate to the designers and investors the impact of their possible choice of retrofit solutions. The toolbox provided a database of refurbishment potential for the houses (Fig. 5.2).

The results showed that insulating the external walls was the most effective building envelope measure. Moreover, we may state that high quality double glazing presented optimal option for windows replacement in Serbia. The case study showed how choosing the measures from the toolbox could enrich the saving effect (Fig. 5.3).

Due to our sampling procedure, the findings could be helpful to learn how to retrofit the houses from the 1970s and 1980s in the most suitable manner. We recommend that house owners and planners comprehend economical and social assessment, and thus chose the appropriate retrofit for a particular house.

Knowing the potential of single measures improves the retrofitting in early design phase. The systematic approach could create a good ground, a pre-step to contribute to the better refurbishment design in future.

Acknowledgments This chapter was realized within the PhD thesis "Patterns for Energy Efficient Design in Serbia" supported with Herder Scholarship by the Alfred Toepfer Stiftung FVS.

References

1. NEUHOFF, K. et al. (2011) *Thermal Efficiency Retrofit of Residential Buildings: The German Experience*. [Online] Climate Policy Initiative Berlin. Available from: http://www.buildup.eu/publications/. [Accessed: 10 May 2013].
2. SERBIAN ENERGY AGENCY. (2013) *Annual Report 2012*. Belgrade.
3. MINISTRY OF MINING AND ENERGY OF THE REPUBLIC OF SERBIA. (2010) *The First Energy Efficiency Plan of the Republic of Serbia for the Period from 2010 to 2012*. Belgrade.

4. BOJIC, M. et al. (2012) Decreasing energy consumption in thermally non-insulated old house via refurbishment. *Energy and Buildings* 54. 503–510.

5. KONSTANTINOU, T., KNAACK, U. (2013) An approach to integrate energy efficiency upgrade into refurbishment design process, applied in two case-study buildings in Northern European climate. *Energy and Buildings* 59. 301–309.

6. MINISTRY OF CONSTRUCTION AND URBAN PLANNING OF THE REPUBLIC OF SERBIA. (2011) *Pravilnik o energetskoj efikasnosti zgrada.* Belgrade: Sluzbeni glasnik RS, br.61/2011.

7. BOINTNER, R. et. al. (2012) *Gebäude maximaler Energieeffizienz mit integrierter erneuerbarer Energieerschließung.* Wien: Bundesministeriums für Verkehr, Innovation und Technologie. Berichte aus Energie-und Umweltforschung 56a/2012.

8. SERBIA. STATISTICAL OFFICE OF THE REPUBLIC OF SERBIA. [Online] Available from: http://www.stat.gov.rs/. [Accessed: 1 May 2013].

9. VOJINOVIC, M. (ed.). (1984) *Katalog projekata.* Edition VII. Belgrade: Naš stan.

Estimating Solar Energy Potential in Buildings on a Global Level

6

Ksenia Petrichenko

Abstract

This chapter contributes to the debate around net-zero energy concept from a global perspective. By means of comprehensive modelling, it analyses how much global building energy consumption could be reduced through utilisation of building-integrated solar energy technologies and energy-efficiency improvements. Valuable insights on the locations and building types, in which it is feasible to achieve a net-zero level of energy performance through solar energy utilisation, are presented in world maps.

6.1 Introduction

Comprehensive scenario analysis for global thermal energy use in the building sector, conducted by the Centre for Climate Change and Sustainable Energy Policy (3CSEP) under the umbrella of Global Energy Assessment (GEA) [1] and Global Building Performance Network (GBPN) [2], has shown that without active and ambitious proliferation of existing energy efficiency best practices in buildings, global thermal energy use will substantially increase (by 32.5 %, according to GEA, and by 48 %, according to GBPN, by 2050 in relation to 2005). At the same time, large-scale proliferation of energy efficiency of new and retrofitted building worldwide will almost halve the global thermal energy use in the building sector [1, 2].

Such a reduction, however, is still not enough for addressing climate change challenge. In order to further reduce emissions from buildings, it is necessary to cover the remaining

K. Petrichenko (✉)
ResearcherCopenhagen Centre on Energy Efficiency (C2E2)UNEP DTU PartnershipMarmorvej 51,2100 Copenhagen Ø, Denmark
e-mail: ksepetrichenko@gmail.com

© Springer Fachmedien Wiesbaden 2015

45

G. Dell, C. Egger (eds.), *World sustainable energy days next 2014,*
DOI 10.1007/978-3-658-04355-1_6

energy needs from renewable energy sources (RES). If building's energy demand is completely covered by renewable energy production, it is usually considered as a net-zero energy building (NZEB) [3].

This chapter is devoted to explore global technical potential for solar energy produced by advanced building-integrated technologies to cover building energy needs, which have been significantly reduced through energy efficiency improvements. The major novelty of the research behind this chapter is the combination of detailed methodology with global coverage.

The research aim of this chapter is to analyse the role of solar-supplied energy-efficient buildings in mitigating climate change from the perspective of bringing the global building sector closer to the net-zero energy (NZE) level.

6.2 Methodology

The chapter is based on a novel methodology, combining bottom-up energy modelling with geospatial analysis. A comprehensive bottom-up energy model 3CSEP-HEB (Centre for Climate Change and Sustainable Energy Policy High Efficiency Buildings) has been developed by the research team (including the author of this chapter) and allows for estimation of the global building thermal energy demand (for space heating, cooling and water heating). In order to evaluate what share of the energy demand can be met by building-integrated solar energy production, a separate model (Building Integrated Solar Energy model—BISE model) has been developed by the author, which takes into account various geographical, architectural, morphological and climatic parameters by means of rigorous geospatial analysis. Combining the results on energy demand and potential solar energy supply provides a valuable insight on the locations and building types where it is feasible to achieve a net-zero level of energy performance through solar energy utilisation.

As the methodology and results of the 3CSEP model have been well documented (see [2, 4–6]), this chapter will describe it very briefly and focus more attention on the estimation of solar energy potential in buildings.

6.2.1 Scope

The analysis has been performed for 11 regions, which cover the globe. The period between 2005 and 2050 is considered as a time frame for the analysis. Data on energy use for space heating, cooling and water heating were taken from 3CSEP-HEB model, while the results for lighting and appliances were provided by Bottom-Up Energy Analysis System (BUENAS) model [7].

In terms of solar energy technologies, the chapter focuses on hybrid photovoltaic/thermal (PV/T) systems. A PV/T system is a combination of photovoltaic (PV) panels and solar thermal components, able to generate both electrical and thermal energy [8]. Such a hybrid installation often has higher performance than separate systems [9].

6.2.2   BISE Model's Design

BISE model assumes that PV/T technologies are being installed in buildings during construction or renovation starting from 2014 and becoming a common practice for all retrofit and new buildings by 2025 in all 11 regions. This process is accompanied by ambitious energy efficiency improvements in buildings worldwide (as in Deep Efficiency Scenario of HEB model).

BISE model uses NASA data for several input parameters (e.g. global solar radiation, top-of-atmosphere solar radiation, ambient temperature, wind speed, etc.) for every hour of 2001–2005, for which average is calculated to get an hourly profile of a typical year [10].

Using data for global and top-of-atmosphere solar radiation, the hourly solar radiation on the (tilted) plane of the solar system is calculated, presenting the amount of solar energy, which can be converted by the technology into heat or electricity. The optimal tilt of the solar system has been assumed. After that, solar electric and thermal energy supplies are calculated separately, according to the formulas often used to calculate the performance of an individual PV/T system. The method presented here for calculating energy collected by one square metre of a solar system per hour has been adapted from [11–14].

All calculations are performed for one square metre of the solar system's surface and then the results are multiplied by the roof area available for solar system installations in each region, climate zone and building type.

This area is calculated by reducing the roof area by several factors, obtained from the literature [15]. Roof area is calculated by multiplying floor area for each region, climate zone, building type, building vintage and year, derived from 3CSEP-HEB model, by roof-to-floor ratios, which were calculated by processing the data sets for urban building areas described in [16] by means of ArcGIS software. Each data set contained the location of different types of built-up areas (low, medium and high) within urban territories. The additional generic Excel data set with a percentage of roof area for each type of built-up densities was also used [17]. For rural areas, the same roof-to-floor ratios were assumed for corresponding building types.

6.3   Results

In this chapter, the results are presented for two building types: single-family and office buildings, for January and July of 2050 separately for thermal and electric energy (see Figs. 6.1 and 6.2).

The results for solar energy are presented as a percentage (from 0 to 100%) of the energy use, illustrating the amount of energy needs, which can be covered by potentially produced solar energy (solar thermal energy is assumed to be used for water heating and space heating, while solar electric energy—for lighting, cooling and appliances) during a certain period of time. Such a comparison provides an insight on in which regions, climate zones, building types and months of a certain year it is possible to achieve NZE goal.

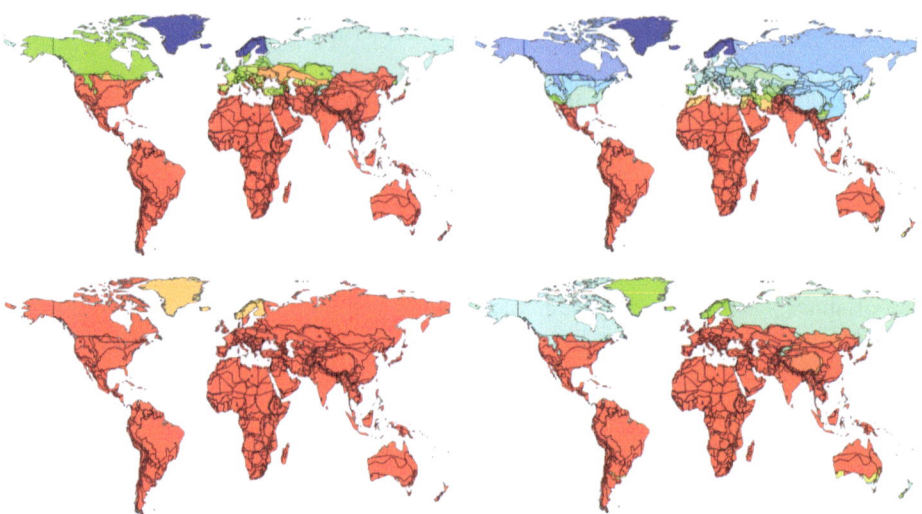

Fig. 6.1 Potential for solar thermal energy to cover thermal energy needs, 2050

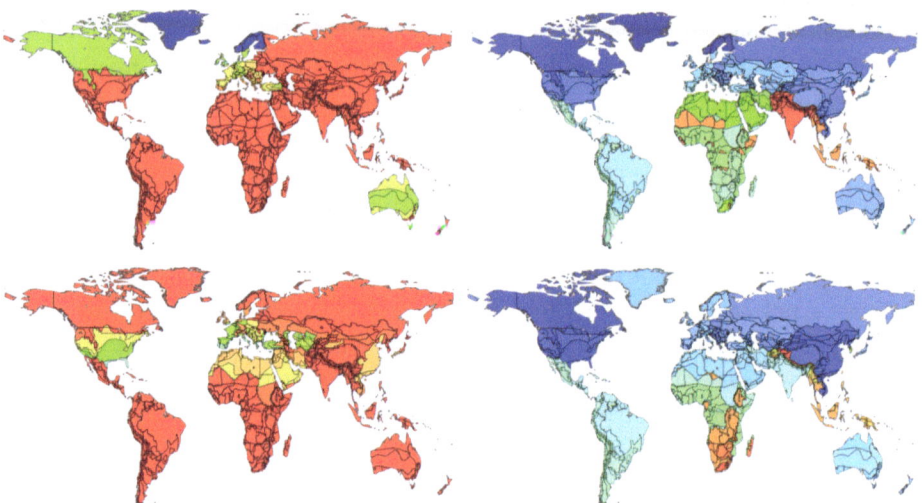

Fig. 6.2 Potential for solar electric energy to cover electric energy needs, 2050

Overall, the results show that solar energy can significantly contribute to the achievement of the NZE goal in buildings located in very different climatic conditions. However, it is likely that in developed countries, growing economies and places with cold climate, building-integrated solar systems have to be combined with other technologies (e.g. heat pumps) in order to generate sufficient amount of renewable energy to cover all the building energy needs, especially during the periods with energy peak demand.

Significant variations in solar energy potential in buildings both for thermal and electric parts can be observed across different months. As for solar thermal energy output, January is the month with the lowest potential (among presented ones) in the Northern Hemisphere, as it is one of the coldest months with limited amount of sunshine and high heating demand.

Electric energy production in 2050 is not sufficient for achieving NZEBs in many locations and for several months. It can be explained by business-as-usual electricity needs from BUENAS model, especially in developed countries and significant needs for cooling during hot months.

The results show that it is not possible to get to net-zero level of energy performance in office buildings by using only solar energy. The main reason for this is that office buildings are much higher than single-family buildings and, therefore, have much smaller roof area available for solar systems' installations in relation to floor area and energy needs. Therefore, the opportunity to produce sufficient amount of solar energy to cover respective energy needs in offices is much more limited than in low-rise buildings.

In case of the solar thermal energy, there is a substantial potential to cover energy needs with solar energy for both building types during the warm season in most of the locations.

For solar electricity, the results are quite different. In single-family buildings, it is possible to achieve 100 % coverage for electric energy needs in a number of locations during the cold season (with the exception for North America and Europe, where energy needs for lighting and appliances assumed in BUENAS model for these regions are significant for all the months). During the summertime, the potential to get to net zero through on-site solar energy generation in single-family buildings decreases in most of developed regions (e.g. North America, Europe, former Soviet Union, Australia, etc.), as relatively high energy needs for lighting and appliances, which do not vary significantly among the months, become accompanied by increased cooling. Office buildings demonstrate much lower potential to cover electricity needs than single-family ones during all the months due to lower available roof area in relation to high energy needs for appliances, lighting and cooling.

The results presented above show that it is very important to analyse the energy balance (i.e. difference between produced and consumed energy in a building) on a monthly basis. While total results for the year might allow for a building to achieve the NZE goal, it does not necessarily mean that energy needs can be satisfied with solar energy every month, due to variations in the availability of renewable energy within the year.

6.4 Conclusion

The chapter has presented a brief overview of the methodology and selected results of a recently developed model—BISE model, which provides the opportunity to estimate technical potential for solar energy production in buildings and its role for transition to the NZE future of the global building sector.

The model has been based on a comprehensive methodology, which combined energy modelling and geospatial analysis, using NASA's hourly climatic data as well as building energy use results from other models, such 3CSEP-HEB and BUENAS models, in order to compare estimated solar energy potential to forecasted energy use for various end-uses in different regions and climate zones.

Advanced visual capacities of the BISE allow for dynamic presentation of the results in the form of the global maps. The results can serve as a scientific basis for the policy-makers, especially when nearly (or NZE) targets are being set in different countries. The results presented in this chapter clearly show the importance of the analysis of the building energy balance for different months, in order to capture the variation in the solar energy potential, which strongly depends on the season and weather conditions. The outputs of the BISE models demonstrated the importance of building types for the discussion of the net-zero concept, as buildings in the same location, but with different configurations, can have very different potentials for achieving NZE status. According to the results, it is unfeasible to attempt satisfying building energy demand only with building-integrated solar energy in high-rise buildings.

Overall, the modelling results show that significant potential for solar energy utilisation exists in most of the building types, regions and climate zones by 2050. If it is efficiently realised on the large scale and combined with wide proliferation of energy efficiency improvements in buildings, it will significantly contribute to the transition to the NZE future.

References

1. ÜRGE-VORSATZ, D. et al. (2012) *Towards sustainable Energy End-Use: Buildings.* In*Global Energy Assessment,* vol. Chapter 10, Laxenburg, Austria, Cambridge, United Kingdom and New York, NY, USA.: IIASA and Cambridge University Press.
2. ÜRGE-VORSATZ, D. et al. (2012) *Best Practice Policies for Low Energy and Carbon Buildings. A Scenario Analysis,* Research report prepared by the Center for Climate Change and Sustainable Policy (3CSEP) for the Global Best Practice Network for Buildings, Budapest, Hungary.
3. ZHU, L. et al. (2009) Comprehensive energy and economic analyses on a zero energy house versus a conventional house, *Energy,* vol. 34, no. 9, 1043–1053, Sep. 2009.
4. ÜRGE-VORSATZ, D., PETRICHENKO, K., BUTCHER, A. (2011) How far can buildings take us in solving climate change? A novel approach to building energy and related emission forecasting. In Energy efficiency first: *The foundation of a low-carbon society,* Belambra Presqu'île de Giens, France.
5. ÜRGE-VORSATZ, D, PETRICHENKO, K. (2013) On the way to the nearly zero future: Results of scenario analysis for building energy use in EU −27, presented at the World Sustainable Energy Days 2013, Wels, Austria.
6. ÜRGE-VORSATZ, D. et al. (2013) Energy use in buildings in a long-term perspective, *Curr. Opin. Environ. Sustain.,* vol. 5, no. 2, 141–151.
7. MCNEIL, M. (2013) Bottom–Up Energy Analysis System (BUENAS)—an international appliance efficiency policy tool, *Energy Effic.,* vol. 5, no. 4, Jan. 2013.
8. DUPEYRAT, P. et al. (2011) Efficient single glazed flat plate photovoltaic–thermal hybrid collector for domestic hot water system," *Sol. Energy,* vol. 85, no. 7, 1457–1468, Jul. 2011.
9. TRIPANAGNOSTOPOULOS, Y. et al. (2002) *Application Aspects of Hubrid PV/T Solar Systems.*
10. NASA (2012) "MDISC Data Subset." Goddard Earth Sciences Data and Information Services Center.

11. DUFFIE, J., BECKMAN, W. (1991) *Solar Engineering of Thermal Processes*, 2nd Edition. John Wiley & Sons.
12. RETSCREEN (2004) Photovoltaic Project Analysis Chapter, in *Clean Energy Project Analysis*, 3rd Edition., Canada: Minister of Natural Resources Canada.
13. RETSCREEN (2004) Solar Water Heating Project Analysis Chapter, in *Clean Energy Project Analysis*, 3rd Edition, Canada: Minister of Natural Resources Canada.
14. IBRAHIM, A. et al. (2009) Performance of Photovoltaic Thermal Collector (PVT) With Different Absorbers Design, *WSEAS Trans. Environ. Dev.*, vol. 5, no. 3, Mar. 2009.
15. IZQUIERDO, S., RODRIGUES, M., FUEYO, N. (2008) A method for estimating the geographical distribution of the available roof surface area for large-scale photovoltaic energy-potential evaluations, *Sol. Energy*, vol. 82 (2008), 929–939.
16. JACKSON, T. et al. (2010) Parameterization od Urban Characteristics for Global Climate Modeling, *Ann. Assoc. Am. Geogr.*, vol. 100, no. 4, 848–865, 2010.
17. FEDDEMA, J. (2011) Personal communication regarding the dataset on the Parameterization of urban characteristics for global climate modeling, 09-Nov −2011.

Long-Term Energy Accumulation in Underground Hot Water Tanks: Fluid Convective Behaviour and Its Influence on the Thermal Losses

7

Milan Rashevski, H. D. Doan and K. Fushinobu

Abstract

Long-term energy storage has great potential to decrease the consumption in buildings, particularly through saving summer excesses to cover winter demands, thus equilibrating the system in both seasons. This chapter deals with hot water storage over 6 months period for an example storage facility in Sofia. Using exact building and weather data in a mathematical model, taking into account fundamental thermodynamical relations, we show amount of energy stored, impact of convection on the thermal losses and we make reference to the building requirements and possible improvements.

7.1 Introduction

Energy consumption and energy saving have become priority concern in the twenty-first century. About 40 % of energy consumption in EU is generated by building consumption [1]—the highest demand compared to the rest of the sectors of the economy. Due to low efficiency of façade solutions and lack of proper architectural design heating and cooling form, about 65 % of this demand [2] is in the occupied residence buildings (including domestic hot water (DHW) and electricity for the stated purposes). One of the possible

M. Rashevski (✉)
Tokyo Institute of Technology, Haidushka gora 147, Sofia 1404, Bulgaria
e-mail: mrashevski@gmail.com

H. D. Doan · K. Fushinobu
Tokyo Institute of Technology
2-12-1 Ookayama, Meguro-ku, 152-8550 Tokyo, Japan
e-mail: mrashevski@hotmail.com

© Springer Fachmedien Wiesbaden 2015
G. Dell, C. Egger (eds.), *World sustainable energy days next 2014*,
DOI 10.1007/978-3-658-04355-1_7

solutions of this problem is to shift the excess of energy in the time when it is needed. While for a short term (e.g. 24-h cycle) there are many working solutions, the long-term storage is a subject that remained open over decades. High technological solutions such as reversible fuel cells or compressed air facilities are not always economically reasonable or even possible to install in urban environment. Furthermore, changing solar thermal energy into a chemical or potential one and getting back to thermal energy for heating purposes in the winter includes great transformation losses.

If we choose not to transform the solar thermal gains, but to store them directly, we should probably use a conventional device such as thermal deposit. However, if we scale it to a building or district size, there are few things that we have to consider. The medium within the deposit should have great volumetric thermal capacity in order to achieve maximum energy stored within the smallest possible volume and consequently to reduce investment costs. In addition, if it is a fluid, its convection within the vessel will cause additional losses due to the hot particles moving close to the boundary layers and causing higher heat flow to the exterior and eventually higher thermal losses. Moreover, increasing the size of the deposit, the velocity of the particles also increases, accelerating the intensity of thermal losses.

Water is one of the compound elements with highest specific thermal heat capacity and easy to use for many practical reasons. As described by Thorsten Urbaneck et al. [3], its strong convection in big vessels is highly problematic and may be reduced by mixing it with gravel. The mixture shows volumetric thermal capacity reduced by approximately 30% and thermal losses dominated by conduction, rather than convection. The initial investment and space requirement in such case is considerably higher with an improvement of exploitation costs, compared to the water-only storage. Simulating the latter one, we are able to describe this improvement in a quantitative way for future design considerations.

7.2   Case Description

Part of the project for the first net-zero energy office building (NZEB) in Bulgaria, designed by Institute for Zero Energy Buildings, Sofia, includes seasonal underground hot water storage connected to solar collectors and floor heating systems. For the simplicity and clarity of the results, we decide to exclude the dynamic solar contribution and dynamic heating consumption, simulate a closed vessel from an initial state, calculate thermal losses through the envelope only and calculate final energy available after the 6-month period.

As seen in Fig. 7.1, the thermal storage contains 288 m^3 of water with initial temperature of 90 °C, insulated by 120-cm top insulation, 80-cm average peripheral insulation and 70-cm bottom insulation of XPS with thermal conductivity of 0.032 W/m^2 K. Specific heat capacity of the water is 4191 J/kg K [4] and thermal conductivity of the water used in the model is 0.66 W/m^2 K [5], both values considered at 70 °C, which is close to the mean temperature over the whole period. Soil thermal conductivity is considered to be 0.6 W/m^2 K and underground temperature of 15 °C below 2 m, values close to the average ones for this region. Soil is heated by the water tank, but conducting rapidly, thus its tempera-

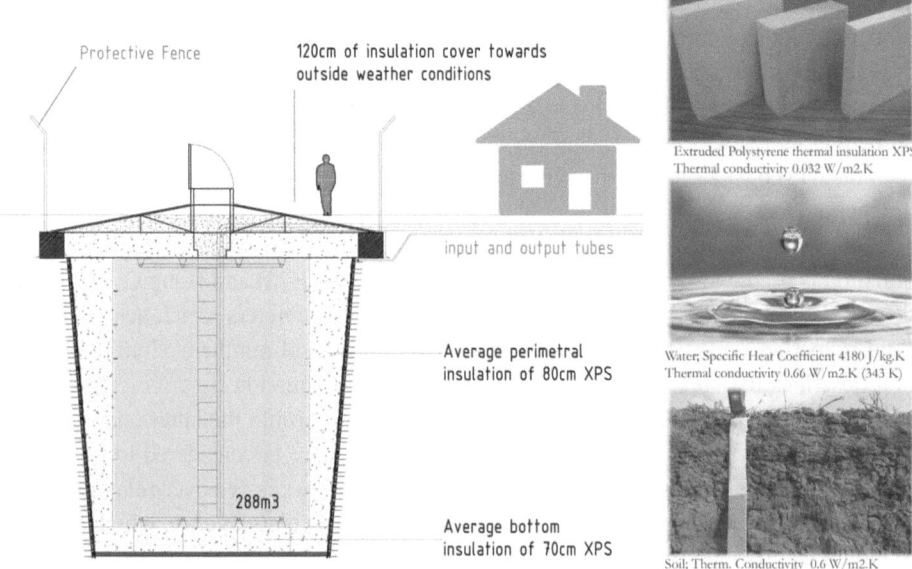

Fig. 7.1 Geometry and materials used

ture is considered to be constant. The insulation layer thickness is increasing, approaching the surface, in order to assure the same heat flow throughout the whole depth of the tank, which is reflected in the model by using constant temperature from 0.25 to 8 m depth. Surface air temperature changes according to the statistical weather data for the period between October 1 and March 31 [6].

The useful water temperature inside the tank should be always defined excluding 40 °C mean floor heating temperature (TH), and thermal energy is defined by $Cp*\rho*V*(T–T$H). In this sense, for the initial state of tank with uniform interior temperature of 90 °C, the initial useful thermal energy is approximately 16.4 mWh.

7.3 Methodology

All the calculations were performed and visualised with Matlab code. Calculations are performed in two space dimensions, assuming X and Y directions being equal due to the symmetrical geometry of the water tank. Finite differences method was used for the discretisation of the differential equations with a domain of 24×32 non-dimensional units, each one corresponding to 25 cm in a real scale. The equations were solved for a period of 6 months with a time step of 30 s. The variables are calculated starting by initial values defined for the first time step. Boundary conditions are updated after each loop and are defined using four temperatures—exterior soil/air temperatures, exterior surface temperature, interior surface temperature and interior neighbour fluid temperature. Similar BC definition is described by R. Vilums [7]. Finally, the model returns to real dimensions and plots the results by means of graphics and important end values.

The first stage executed was to calculate conduction only through the temperature equation and thermal diffusivity as parameter. A separate formulation described by Incropera et al. [5] was used to confirm the results. This calculation allows a comparison with the one considering laminar flow and first evaluation of the convection influence on the thermal losses.

For the calculation of the convection, a non-dimensional vorticity–stream function formulation is used, close to the one described by C. Shu et al. [8] and Hong G. Im [9] with three differential equations to be solved and two main parameters which describe all the fluid properties and vessel geometry—Rayleigh and Prandtl numbers. Continuity equation describes the stream function, advection–diffusion equation describes the vorticity and temperature equation allows us to input the buoyant term in the latter one. The model is particularly suitable for laminar flows, where pressure is considered to be uniform throughout the vessel and no pressure terms are needed. Successive over-relaxation iteration method is used to solve the continuity equation for fast convergence, which allows us to reduce significantly the calculation time [10].

The definition of the Rayleigh and Prandtl numbers shows that the value of the first one is far below the one we should use for the case 4×10^{14}. The one in the model is 1×10^7, and each higher value results in an unstable model. If we go deeper in the significance of Rayleigh number, we see that it is the product of Prandtl and Grashof numbers, giving information about the convection/conduction ratio, responsible for the heat transfer in buoyancy-driven flow, and particularly about the dimensions of the vessel, expansion characteristics of the fluid, temperature difference and associated diffusivities. The high value of our case allows us to conclude that the acceleration of the particles for 8 m of height and the temperature differences are so great that the model is approaching transitory turbulent flow. This consideration leads us to the far more three-dimensional calculations due to the three-dimensional nature of the turbulence. Furthermore, pressure term could not be neglected anymore and additional equations are needed to describe it. For this particular model, without the need to go deep into details of eddies geometry, the simplest turbulent k–e model was studied, using two additional equations for the pressure term and RANS [11]. Two more variables—kinetic energy "k" and dissipation length "e" are included, and turbulent Prandtl number were analysed for the case [12, 13].

7.4 Results

The results from conduction-only calculations show a temperature drop down to approximately 81.5 °C. The temperature difference within the vessel is about 13 °C (as shown on Fig. 7.2) and the thermal losses through the envelope are minimal, result of pure conduction process, if water is considered to be a solid. Final thermal energy available is 13.6 mWh, losses are about 2.8 mWh, and considering the initial energy of 16.4, gives us an overall efficiency for the period of 83 %.

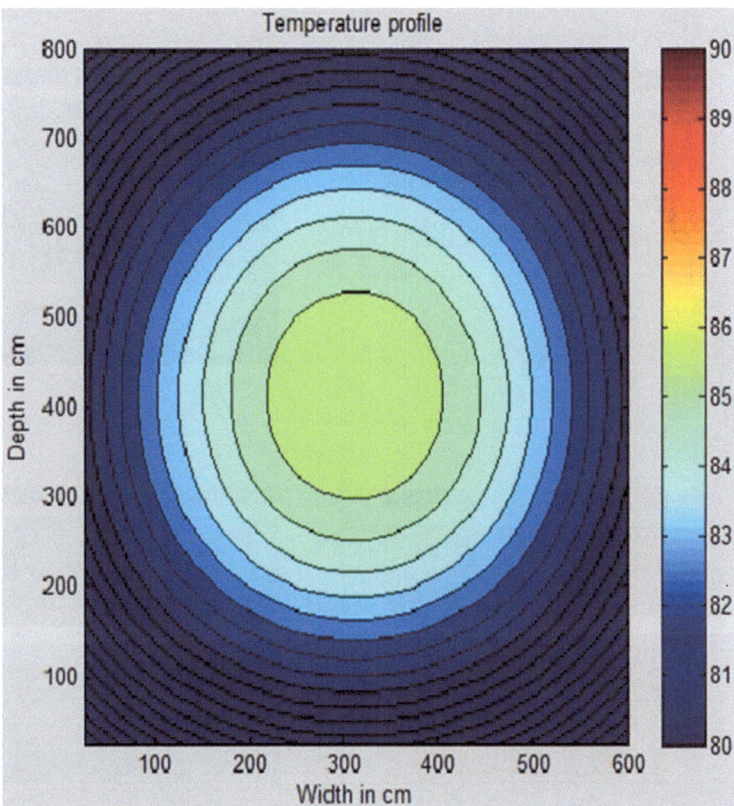

Fig. 7.2 Temperature profile after 6 months for conduction model

Considering laminar flow with the maximum R_a value allowed for the stability of the model, the results show a temperature drop down to 63.8 °C, useful thermal energy of 7.8 mWh and an overall efficiency of 48 % (Table 7.1). The need for turbulent model to describe the real situation is further confirmed by high water velocity, obvious mixing and homogeneity of the temperature with differences up to 2 °C for the whole 288 m³ water volume (Fig. 7.3).

7.5   Conclusions

Conduction influence for the thermal losses was proven in a quantitative way, showing its great impact on the decrease of the water temperature and useful thermal energy. In the case studied, the temperature decrease from conduction only is 8.5 °C, while for conduction and convection, it is more than 26 °C. Furthermore, there is almost no temperature stratification of the water which does not allow using hot water for longer periods, when the average temperature of the water tank is lower than the required one. The efficiency drops from 83 down to 48 %, which in terms of energy means 5.8 mWh. (Table 7.2)

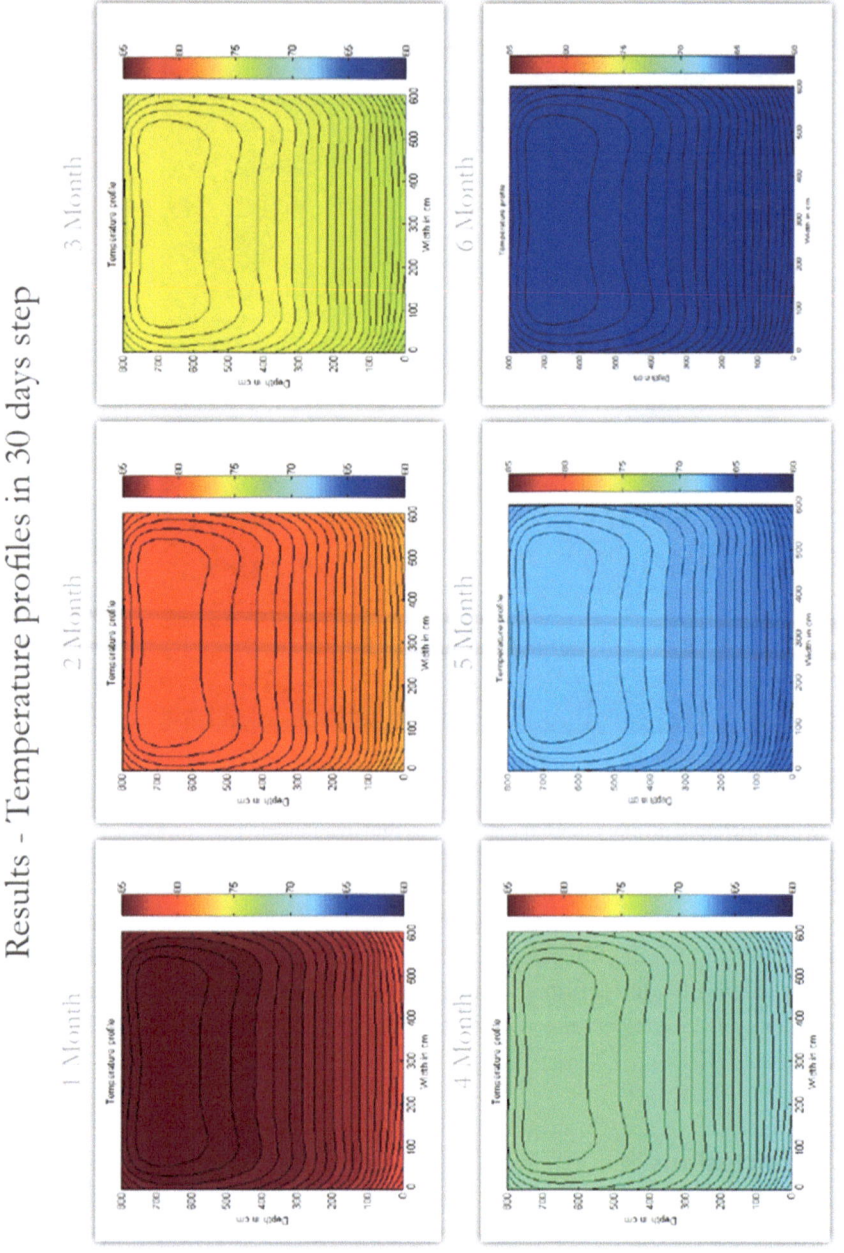

Fig. 7.3 Temperature profiles after each month for laminar flow

Table 7.1 Temperature and energy results after each month for laminar flow

Period	Max Temp (T°C)	Aver Temp (T°C)	Min Temp (T°C)	Temp Differ (T°C)	Disponible Energy (mWh)	Energy Losses (mWh)
Initial state	90.000	90.000	90.000	0.000	16.395	0.000
1 Month	85.069	84.284	82.244	2.825	**14.521**	1.874
2 Months	80.313	79.584	77.685	2.628	12.980	1.541
3 Months	75.883	75.207	73.442	2.441	**11.544**	1.435
4 Months	71.756	71.129	69.489	2.267	10.207	1.337
5 Months	67.911	67.330	65.806	2.105	8.962	1.246
6 Months	64.327	63.788	62.372	1.955	7.80 mWh	1.161
Efficiency of the system for an year cycle:					48%	8.595

Table 7.2 Influence of water movement intensity on the thermal losses

Behavior	Max Temp (T°C)	Aver Temp (T°C)	Min Temp (T°C)	Temp Differ (T°C)	Disponible Energy (mWh)	Energy Losses (mWh)	Efficiency (%)
Initial Energy		90.000			16.4 mWh		
Conduction	86.039	81.455	73.155	12.884	13.6 mWh	2.802	83%
Convection	64.327	63.788	62.372	1.955	7.80 mWh	8.595	48%
Turbulence	???	? 59 ?	???	???	6.23 mWh	???	38%

This significant reduction of the efficiency and lack of temperature stratification make further turbulent model studies for this case of little practical interest. The right approach to choose is to develop methods for reduction of the convection inside the water tank, maintaining the same volumetric heat capacity and similar investment costs. The simplest way suggested is to divide the volume geometry into small cells connected to each other, with a cheap and durable material. Furthermore, in order to cover 100% of the heating demand, one should achieve very ambitious energy-efficient standard of NZEB, or use more volume for less efficient building designs (Table 7.3).

Table 7.3 Consumption rates for heating referred to the available thermal energy

Energy recapitulation balance for the office building ZEB1

Sofia, Bozhurishte; Rehabilitation including the summilated 288m3 thermal accumulator

Available energy (kWh)	Design Surface (m2)	Maximal consumption (kWh/m2a)
7800	1150	**6.8** kWh/m2a

Disponible energy per m3 after 180d (kWh)				**27**
	Consumption (kWh/m2a)	Usable surface (m2)	Required energy for 1 year (mWh)	Required volume (m3)
NearZeroEnergy house	**6**	1150	6.90	256
Passive house	**15**	1150	17.25	639
Low Energy house	**40**	1150	46.00	1704

Available energy in the hot water tank after 180d (mWh)		**7.800**
	Consumption kWh/m2a	Max Usable Surface (m2)
NearZeroEnergy house	**6**	1300
Passive house	**15**	520
Low Energy house	**40**	195

References

1. EUROPEAN PARLIAMENT AND COUNCIL (2010) *Directive 2010/31/EU 19 May 2010 on the energy performance of buildings.* Strasbourg: Official Journal of the European Union (L 153/13)
2. U.S. ENERGY INFORMATION ADMINISTRATION (2009) *Residential Energy Consumption Survey.* [Online]. Available from: http://www.eia.gov/consumption/residential/.
3. URBANECK, T., PLATZER, B. and SCHIRMER, U. (2003) Advanced monitoring of water gravel system, *Futurestock 2003, 9th International Conference on Thermal Energy Storage, Warschau (Polen),* Proceedings Vol.1, S. 451-458, ISBN 83-7207-435-6
4. THE ENGINEERING TOOLBOX (n.d.) *Thermal properties of the water* [online] Available from: http://www.engineeringtoolbox.com/water-thermal-properties-d_162.html.
5. INCROPERA, F.P. et al. (2011) *Heat and Mass Transfer,* 6th Ed. Hoboken: John Wiley & Sons, Inc., ISBN 13 978-0470-50196-2
6. U.S. DEPARTMENT OF ENERGY (n.d.) Energy+Weather Data [Online] Available from: http://apps1.eere.energy.gov/buildings/energyplus/weatherdata_about.cfm
7. R. VILUMS (2011) *Implementation of Transient Robin Boundary Conditions in OpenFOAM,* Riga: ESF Project No. 2009/0223/1DP/1.1.1.2.0/APIA/VIAA/008, University of Latvia, Faculty of Physics and Mathematics.
8. SHU, C. and YEO, K.S (2000), An Efficient Approach to Simulate Natural Convection in Arbitrarily Eccentric Annuli by Vorticity-Stream Function Formulation, *Numerical Heat Transfer,* Part A, 38: Pg. 739-756, Copyright: Taylor & Francis
9. HONG G. IM (2001) A Finite Difference Code for the Navier-Stokes Equations in Vorticity Stream Function Formulation, University of Michigan, [Online] Available from: http://www.fem.unicamp.br/~phoenics/SITE_PHOENICS/Apostilas/CFD-1_U%20Michigan_Hong/Lecture05.pdf
10. GIORDANO N.J. and NAKANISHI H. (2012) *Computational Physics using MATLAB,* ISBN: 0-13-146990-8, 2nd Edition, West Lafayette: Prentice Hall

11. PATANKAR, S.V. (1980) *Numerical Heat Transfer and Fluid Flow,* ISBN: 0-07-048740-5, New York: McGraw Hill Book Company.
12. FERTZIGER J.H., and PERIC M. (2002), *Computational Methods for Fluid Dynamics,* ISBN 3-540-42074-6, 3rd Edition, Berlin; Heidelberg; New York; Barcelona; Hong Kong; London; Milan; Paris; Tokyo: Springer
13. HASAN B.O. (2007) Turbulent Prandtl Number and its Use in Prediction of Heat Transfer Coefficient for Liquids, *Nahrain University, College of Engineering Journal (NUCEJ),* Vol.10, No.1, Pg.53-64.

Optimizing the Control Strategy of a Low-Energy House's Heating System

8

Matteo Rimoldi, Elisa Carlon, Markus Schwarz, Laszlo Golicza, Vijay Kumar Verma, Christoph Schmidl and Walter Haslinger

Abstract

The objective of this work is to analyze the heat demand of a lowenergy house and to optimize the heating system control strategy. Low-energy houses minimize their heat losses by means of a highly insulated envelope. The considered heating system is composed of a pellet boiler supplying hot water to a floor heating system. The annual heat demand has been calculated with a focus on the effect of internal heat gains. Different solutions have been investigated to optimize the control strategy.

8.1 Introduction

Residential buildings contribute to more than 35 % of the final energy consumption, and to a large part of the green house gas (GHG) emissions in the EU [1]. In 2010, the EU adopted a Directive on the energy performance of buildings, to reduce the energy consumption of the building sector [2]. Houses having an annual heat demand below 50 kWh/m^2 are considered as low-energy houses according to the Austrian regulations [3, 4]. For

M. Rimoldi (✉) · E. Carlon · M. Schwarz · L. Golicza · V. Verma · C. Schmidl · W. Haslinger
Bioenergy2020+GmbH, Gewerbepark Haag 3, 3250 Wieselburg, Austria
e-mail: matteo3.rimoldi@mail.polimi.it

M. Rimoldi
Politecnico di Milano, Piazza Leonardo da Vinci 32, 20133 Milano, Italy

E. Carlon
Free University of Bozen–Bolzano, Universitätsplatz—Piazza Università 5,
39100 Bolzano, Italy

© Springer Fachmedien Wiesbaden 2015
G. Dell, C. Egger (eds.), *World sustainable energy days next 2014*,
DOI 10.1007/978-3-658-04355-1_8

a biomass boiler, the amount of CO_2 emitted during combustion is the same that was absorbed from the atmosphere during the biomass growth; therefore, there is no addition of CO_2 to the environment. The aim of this work is to analyze the energy performance of a low-energy house and to optimize the heating system control strategy. First, the annual heat demand has been calculated by means of a dynamic simulation tool, evidencing the contributions of building envelope and internal heat gains. Second, a hydronic heating system, equipped with a biomass boiler, has been implemented in the simulation. This study has been performed on a low-energy house currently monitored in the frame of the BioMaxEff project [5], an FP7 project aiming at the demonstration of ultralow emission and highest efficiency performance biomass boilers in real-life operating conditions.

8.2 Case Study

The house analyzed in this work is a low-energy prefabricated house built in 2012. Its lightweight external walls have thick layers of insulating materials, and all the windows are triple glazed. A considerable fenestrated surface on the southern facade maximizes the solar radiation gains. Moreover, part of the energy required to heat the house comes from the house itself: people, appliances, and lights are heat sources that contribute to the global energy balance. The house is heated by means of a floor heating system, composed of a network of plastic pipes embedded in a concrete layer. Hot water flowing through the pipes increases the temperature of the surrounding materials, thus transferring heat to the rooms. In a floor heating system, an inlet temperature between 25 and 40 °C ensures a good heat exchange through the floor without exceeding international standards limitation for floor temperature, stated to be lower than 29 °C [6]. This temperature is achieved by mixing the hot water from the boiler with the return flow from the floor heating system. The boiler is a 6 kW pellet boiler manufactured by an Austrian company. Its basic unit consists of a steel boiler with a speed-controlled vacuum fan, a cyclic pellet metering auger, and a microprocessor that controls the combustion process.

8.3 Method

The energy performance of house has been analyzed by means of a dynamic simulation software (Energy Plus) [7]. The heated volume of the house (544.82 m^3) has been divided into four different thermal zones (two for each floor) in order to analyze in detail the influence of different orientations (two zones are south exposed and two zones are north exposed) combined with a wider fenestration surface.

8.3.1 Heat Gains

Internal heat sources are represented by people, lights, and electric appliances. These heat sources are highly influenced by the lifestyle of the people who live in the house.

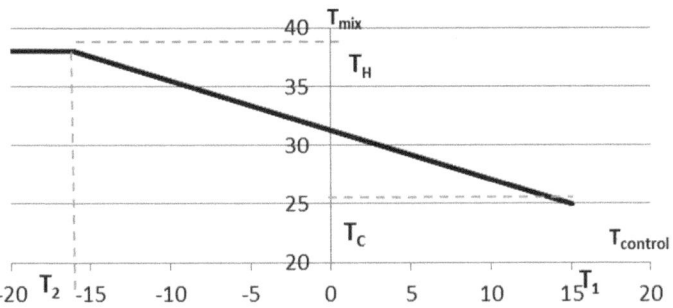

Fig. 8.1 General control scheme

Information about lifestyle, type, and location of lights and appliances was provided by a questionnaire to the house owner.

8.3.2 Pellet Boiler

The inputs for the boiler model available in Energy Plus are the boiler's nominal capacity and performance curve. This curve represents the boiler efficiency as a function of the partial load ratio (PLR). For the considered pellet boiler, the curve has been obtained based on experimental data registered during tests performed in the Bioenergy 2020+technical laboratory (48.117°N 15.136°E, 270 m above mean sea level)[8]. A 6-kW pellet boiler was tested at different loads and its performance curve was determined with a least square regression on the test results. The equation of the curve is:

$$\eta_{LHV} = -1.581 \cdot PLR^2 + 2.205 \cdot PLR + 0.129 \tag{8.1}$$

8.3.3 Floor Heating

The hot water leaving the boiler is mixed with the return cold water to reach a certain temperature (T_{mix}), at which the water is delivered to the pipes of the floor heating. The value of T_{mix} is calculated as a function of a generic control temperature, as shown in Fig. 8.1. When $T_{control}$ goes below the limit value T_1, the heating system is turned on, and the floor heating starts to warm up the house. T_{mix} was assumed to vary in the range of 25–38 °C, as specified by the house manufacturer; therefore, the slope of the line is determined by the choice of T_1 and T_2. The control parameter can be the outdoor temperature or the indoor temperature:

- If the system is controlled by the outdoor temperature ($T_{control} = T_{outdoor}$), T_2 has been fixed at −16 °C, the outdoor design temperature according to the European standard EN 12831[9]. The value of T_1 was optimized according to the simulation results.
- If the system is controlled by the indoor temperature ($T_{control} = T_{indoor}$), both the values of T_1 and T_2 were optimized.

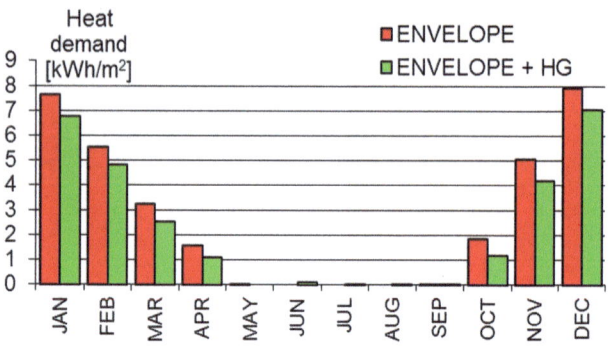

Fig. 8.2 Heat demand considering only envelope (*red*) and considering also internal heat gains (*green*)

8.4 Results and Discussion

8.4.1 Heat Demand

The effect of internal heat gains on the annual heat demand of the house has been evidenced by means of two different simulations: The first simulation considers only the envelope of the house; the second one includes also the contribution of the internal heat gains (Fig. 8.2).

8.4.2 Control Optimization

The heating system control strategy has been optimized for the duration of the whole heating season. The criteria to choose the best solution were the fuel consumption and the number of hours during which an indoor comfortable temperature is ensured (comfort temperature range is 20–27 °C during the day and 18–27 °C during the night). For the control strategy based on the outdoor temperature, three different values of T_1 (the outdoor temperature at which the system starts operation) have been implemented in the simulations, as shown in Fig. 8.3.

$T_{outdoor}$ [°C]		
	T_1 [°C]	T_2 [°C]
OPTION 1	5.0	-16.0
OPTION 2	7.5	-16.0
OPTION 3	10.0	-16.0

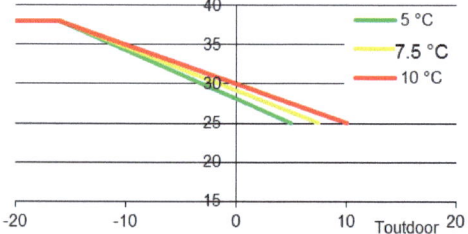

Fig. 8.3 Settings for the outdoor temperature-based control

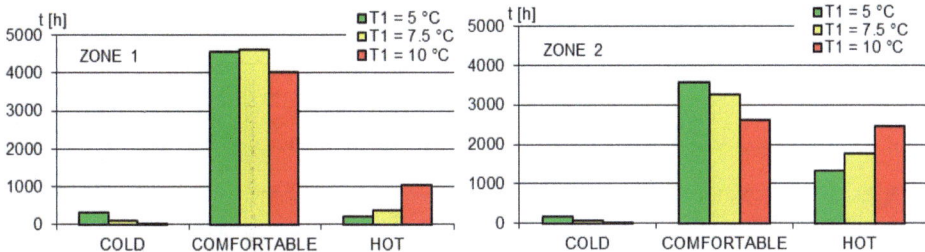

Fig. 8.4 Amount of comfortable hours for zone 1 and zone 2

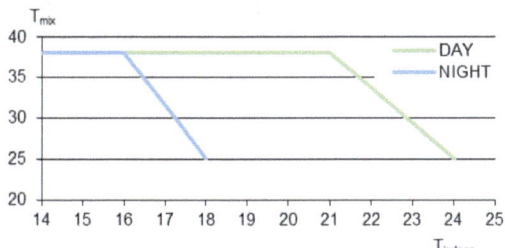

T_{indoor} [°C]		
	T_1 [°C]	T_2 [°C]
Day 05:00 – 21:00	24	21
Night 21:00 – 05:00	18	16

Fig. 8.5 Settings for indoor temperature-based control

Results concerning comfort conditions and fuel consumption are reported in Fig. 8.4 and Table 8.1: the hours in which the room temperature is in the comfort range are indicated in the category "COMFORTABLE." If the room temperature is above or below the comfort range, the hours are classified as "hot" or "cold," respectively.

For the indoor temperature control, different temperatures have been set for daytime and night, as reported in Fig. 8.5.

Table 8.1 Fuel consumption with outdoor temperature-based control. $LHV_{pellet} = 17.17$ MJ/kg (w.b.)

Option	Fuel consumption
$T_1 = 5°C$	1899
$T_1 = 7.5°C$	2139
$T_1 = 10°C$	2301

Table 8.2 Comparison of fuel consumption for outdoor and indoor control

Option	Fuel consumption (kg)
Outdoor temperature-based control	1899
Indoor temperature-based control	1849

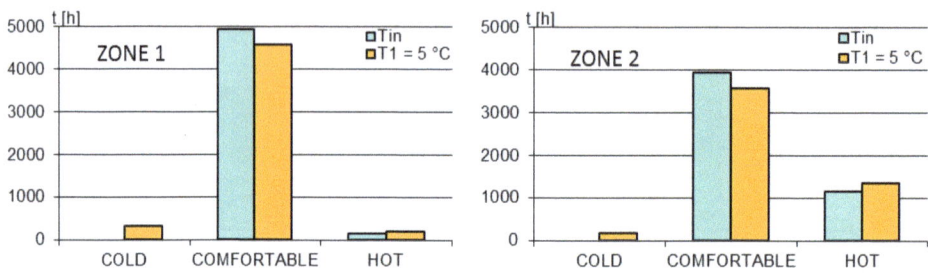

Fig. 8.6 Comparison between comfortable hours for zone 1 and zone 2

The results obtained with the control based on the indoor temperature have been compared (Fig. 8.6 and Table 8.2) with the results of the best outdoor temperature control strategy ($T_1 = 5\,°C$).

8.5   Conclusion and Future Work

In this work, the energy performance of an existing low-energy house has been analyzed and, with the integration of a hydronic heating system, several solutions for the heating system control strategy have been investigated. The heat demand analysis showed that, for a highly insulated house, the influence of internal gains on the heat demand is relevant. As the building requires a low amount of energy ($< 35\,kWh/m^2$), the fraction provided by heat gains becomes more important and reduces by 16% share of heat that must be supplied by the heating system. Then two different control strategies have been analyzed. For the outdoor temperature-based control, several options have been considered with different values of T_1, the base point of the heating curve. Due to the high thermal performance of the envelope, the best choice for the outdoor temperature-based control was the one with the lower value of T_1, ensuring the highest amount of comfortable hours, a reduction of the "hot" hours (see Sect. 8.4.2) compared to the other solutions and a fuel saving of more than 10%. This option has been compared to another control strategy depending on the indoor temperature. Results show that the limit values of $T_{control}$ need to be accurately optimized, to exploit the thermal performance of this building typology, granting a high amount of comfortable hours (as defined in Sect. 8.4.2) and low fuel consumption. The next steps of this work could be represented by further investigation of different solutions for the control system: a combined control strategy considering outdoor and indoor temperature could increase the indoor thermal comfort and reduce the fuel consumption. Furthermore, to avoid overheating in the south-exposed zones, the control strategy could be arranged with different settings, depending on the orientation and the solar radiation incoming in the zones.

Acknowledgment This study has been performed in the frame of the BioMaxEff project [5]. The research leading to these results has received funding from the European Union Seventh Framework Programme (FP7/2007-2013) under Grant Agreement no. 268217.

References

1. PINEAU, D. et al. (2013) Performance analysis of heating systems for low energy houses. *Energy and Buildings*. 65. 45–54.
2. EUROPEAN COMMISSION (2010) EU-Directive 2010/31/EC of the European Parliament and of the council of 19 May 2010 on the energy performance of buildings (recast).
3. THOMSEN, K.E., WITTCHEN, K.B. (2008) European national strategies to move. SBI (Danish Building Research Institute).
4. LAND OBERÖSTERREICH (2013) [Online], Available from: http://www.land-oberoesterreich.gv.at/. [Accessed 20 June 2013].
5. BIOMAXEFF (2013) [Online] Available from: http://www.biomaxeff.eu. [Accessed 1 June 2013].
6. EN 1264–3 (2009) Water based surface embedded heating and cooling systems:
7. ENERGYPLUS (2013) [Online] Available from: http://apps1.eere.energy.gov/buildings/energy-plus. [Accessed 10 July 2013].
8. BIOENERGY 2020+. (2013) [Online] Available from: http://www.bioenergy2020.eu. [Accessed 25 May 2013].
9. EN 12831 (2006) Heating systems in buildings – Method for calculation of the design heat load.

Optimizing Self-Consumption of Grid-Connected PV/Storage Systems

9

Theresa Wohlmuth, Franz Jetzinger and Johannes Schmid

Abstract

An electrical storage system is mainly used to increase selfconsumption of the produced photovoltaic (PV) energy, to relieve the public power grid and to reduce the dependency on the grid. This chapter focuses on a technical simulation of a PV system linked to a storage unit and analyses its economic efficiency.

9.1 Introduction

Electricity is accepted as one of the driving forces for economic development around the whole world. Nowadays, the human population depends on fossil fuels like oil, gas and coal for electricity production. Two major disadvantages are the depletability of these fossil resources and the resulting carbon dioxide emissions when burning them.

To reduce the energy demand and fossil fuels, the following core actions need to take place in the order shown below:

- Energy savings
- Energy efficiency
- Supply of renewable energy sources

T. Wohlmuth (✉) · F. Jetzinger · J. Schmid
ALPINE ENERGIE Österreich GmbH, Winetzhammerstraße 6, 4030 Linz, Austria
e-mail: theresa-wohlmuth@gmx.at

© Springer Fachmedien Wiesbaden 2015
G. Dell, C. Egger (eds.), *World sustainable energy days next 2014,*
DOI 10.1007/978-3-658-04355-1_9

In the course of time, the subsidies for private customers have been decreasing and the costs for electricity per kilowatt hour have been continuously rising. As a consequence, especially grid-connected photovoltaic (PV) systems increased substantially over the past few years [1], resulting in commercial sales of energy to the distribution network if there is any surplus of PV energy. To make these systems economically viable without funding, it is necessary to use as much self-generated PV electricity [2] as possible by the directly connected consumers themselves – this factor is described by the self-consumption (SC) rate [3].

9.2   The Integration of a Storage System

The proper sizing and the correct configuration of the individual parts of the home power plant (load, solar yield and storage) are crucial for the cost-efficient operation [4]. In this context, it is important not to oversize the storage system because the investment costs would be enormous and subsequently the system would not be profitable.

The lifetime can be maximized by the correct operational management of the system. One important point of this consideration is the storage strategy. For the simulation, a simple model has been designed:

- Energy is supplied by the PV system and there is an energy demand in the household, the energy is used by the household itself.
- If the energy produced by the PV system is higher than the currently required demand of the household, the surplus is stored.
- If more power is required than provided by the PV system at the moment, energy is taken from the battery.
- Energy is only taken from the public power grid if the demand of the household cannot be met by the storage system.
- If there is neither a demand nor a free storage capacity, the surplus is fed into the public power grid.

This model for charging and discharging the battery is not a strategy in the proper sense. Storage strategies, for example, take into account weather forecasts and foreseeable loads to properly manage available capacity in the storage. Furthermore, the addition of a charge control, an overcharge protection and a deep discharge protection is important to increase the lifetime of the electrical storage system [5].

One of the reasons why it will be necessary to implement a storage strategy is shown in Fig. 9.1. The frequency distribution of the storage state (empty of full) is recorded for each month with the associated percentage of the average state of charge. According to the simulation in January, the battery is empty 81 % of the time.

This means the capacity sinks to the maximum depth of discharge (DoD) and the battery just works 10 % of the time in the first month of the year. Considering the months

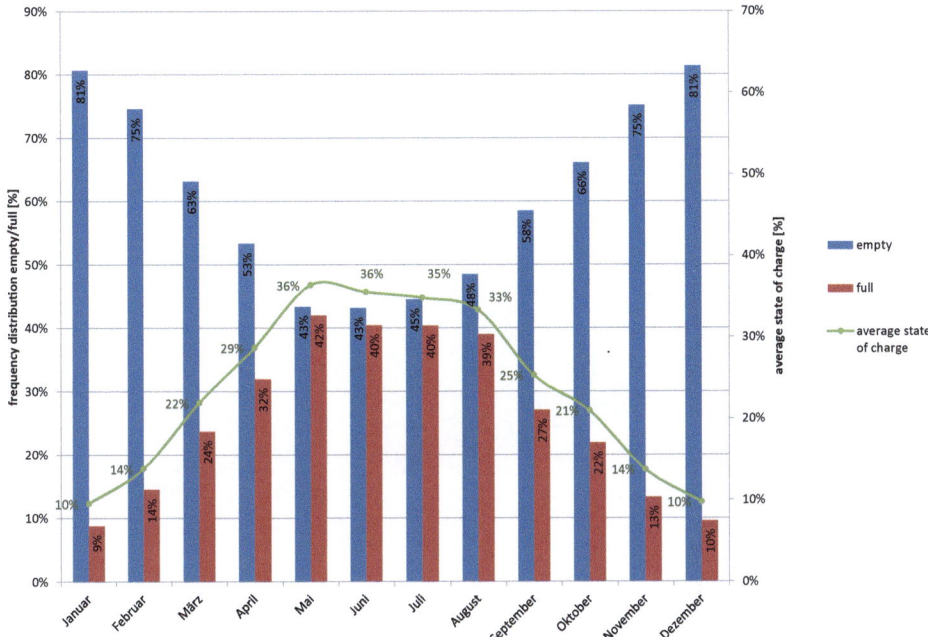

Fig. 9.1 Frequency distribution of empty and full in a 2-kWh storage in comparison with a 5-kWp PV plant. *PV* photovoltaic

May to July, the two bars are almost equal. The battery is full 40% of the time in summer because of the high solar yield, and therefore the working time is also reduced.

In a theoretically perfect operation, the two bars would approach zero. In reality, this cannot be reached: a small battery in winter is empty most of the time and full in summer. This is because of the low capacity of the battery, and thus it is quickly loaded, respectively unloaded.

In comparison, a big storage system is often full and hardly empty in summer (kind of seasonal storage). This damages the battery because there is no full load cycle over a longer time.

9.3   PV/Storage Simulation

The available capacity of the battery consists of the DoD, the maximum charge and the efficiency of the battery. Within the course of the simulation, the charge and discharge efficiencies are not taken into consideration. The cycle stability is closely linked to the lifetime of a battery. This value is significant for the annual complete charge cycles. The size of the exemplary household is based on data from "STATISTIK AUSTRIA" [6] and is rounded to 4000 kWh/a.

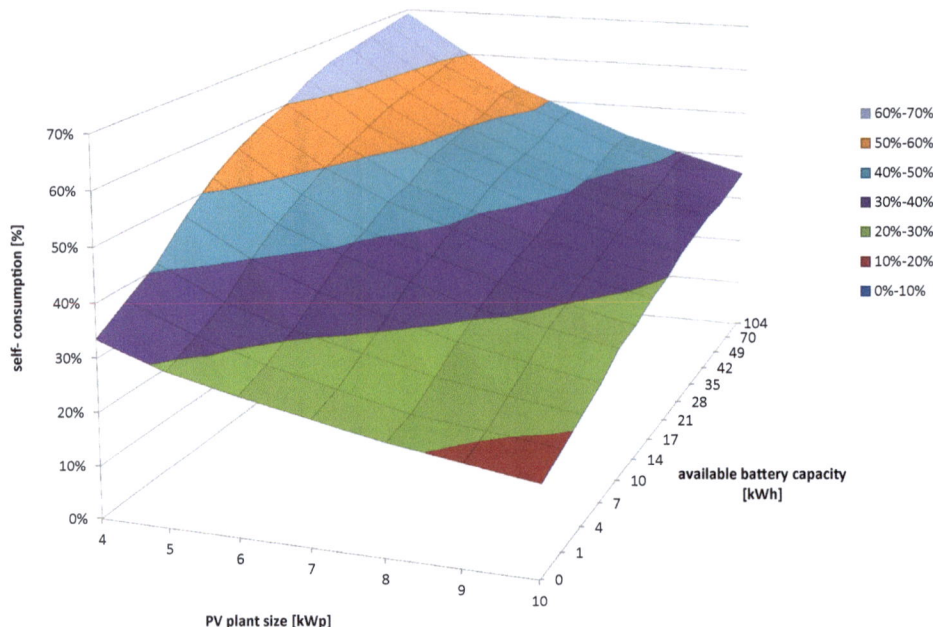

Fig. 9.2 Self-consumption in case of the combination of PV and storage system. *PV* photovoltaic

The same principle is used to calculate the SC rate with an integrated battery system. Figure 9.2 shows the SC rate by varying the PV plant size and the available battery capacity.

The higher the battery capacity, the greater the SC rate because more surplus energy can be stored. The sum of the SC rate and the amount of energy fed into the public grid and also the sum of the solar coverage rate and the energy obtained from the public grid always equal 100%.

It might appear that an increase in battery capacity will continuously lead a proportional increase in the SC rate. However, Fig. 9.3 clearly shows that at a particular value, the increase of the curve slows down. Therefore, a supersized storage would be necessary to achieve 100% of SC rate.

Pros and cons must be carefully considered when choosing the size of the electric storage system. Small batteries have the advantage of being less expensive, but on the other hand, they cannot store the midday peak energy input of a sunny day and cover the evening demand of a household because of the low capacity.

To choose a suitable storage system, all major factors need to be taken into consideration. These include the household with the specific load profile, the PV yield depending on the location, the direction of the PV modules and the efficiency of the inverter and the storage system. Only with a suitable storage system, the SC rate can be increased significantly, and consequently, the profitability of the system is given.

In order to achieve a higher rate of SC, higher storage capacities are needed. In this case, the system will not be amortised in the lifespan of the PV power plant, assuming real

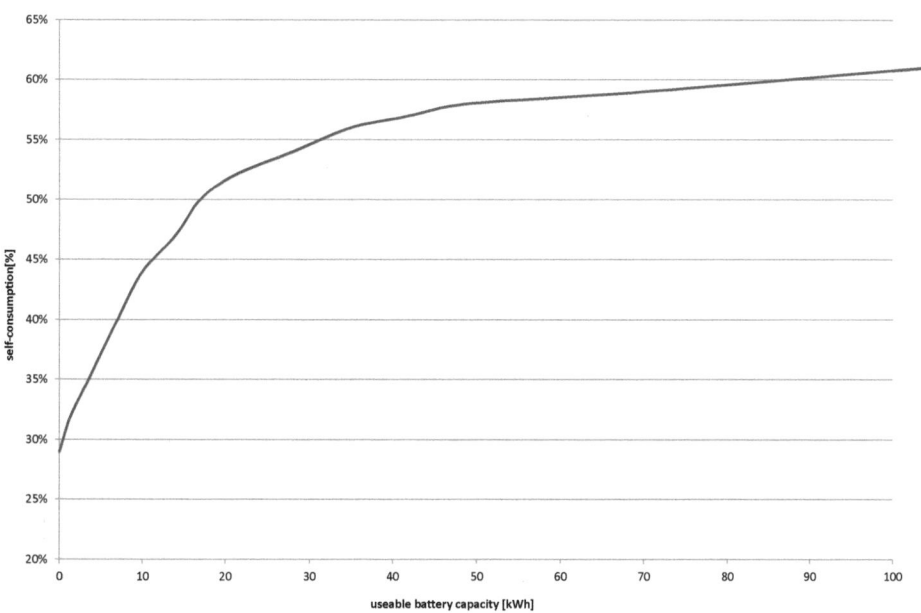

Fig. 9.3 Self-consumption in the case of variable battery and an exemplary household (4000 kWh/a) with a 5-kWp PV. *PV* photovoltaic

specific costs of a battery system of at least 300 €/kWh for lead batteries and 500–1000 €/kWh for lithium batteries [7].

9.4   Economic Assessment

As a basis for the economic calculation of the exemplary household, a storage system with a 5 kWh useable battery capacity (including DoD and the efficiency of the system) was used. For the simulation, network costs of 22 €ct/kWh, a feed-in tariff of 5.5 €ct/kWh and an estimated 2% increase in electricity prices per year [8] are assumed.

The lithium battery with specific costs of 500 €/kWh is not amortised within 25 years. By only changing the increase in electricity prices per year to 4%, the complete system is being amortised within 22 years. This means even a small change in the underlying assumptions causes a significant change in the results.

The result of the economic assessment of an electrical storage system is shown in Fig. 9.4. A lithium battery with a varying capacity between 2 and 150 kWh in combination with a PV plant size of 5 kWp and an annual electricity consumption of 4000 kWh/a is illustrated. The green area in the left corner highlights an amortisation time of more than 25 years.

This calculation does not include subsidies and new storage acquisition. In reality, the storage system should be replaced every 5–15 years due to capacity losses depending on

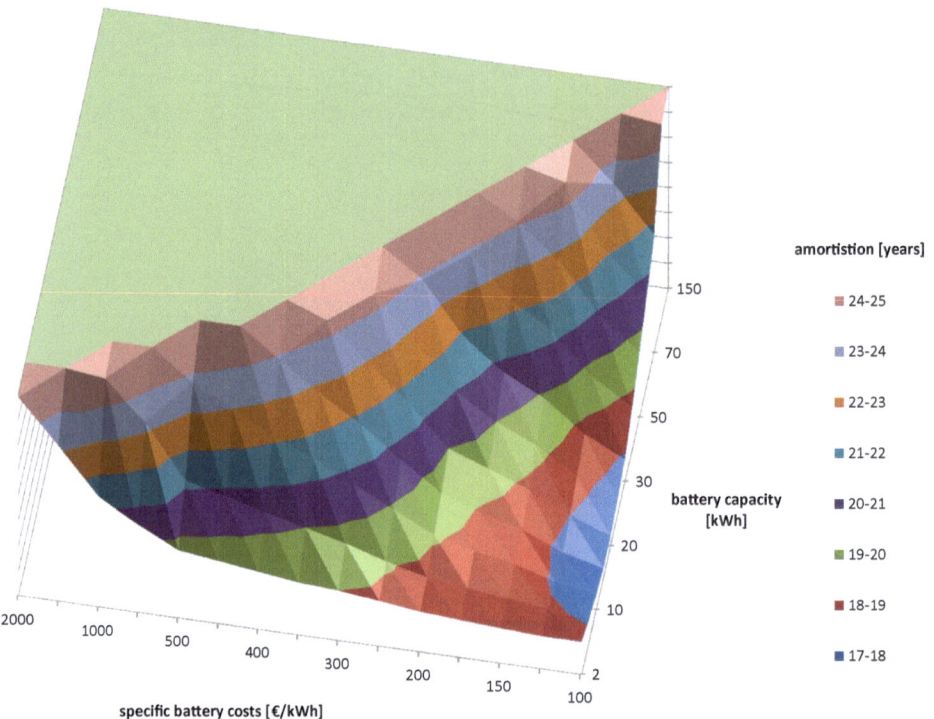

Fig. 9.4 Amortisation of a lithium battery in combination with 5-kWp PV and 4000 kWh/a. *PV* photovoltaic

the storage technology and the operation of the system. This means, based on a typical 25-year lifespan of the PV system, the storage system has to be renewed three to five times.

Compared with the specific grid costs per kilowatt hour shown in Fig. 9.5 the electricity costs for combined PV and battery the point of intersection is reached after 24 years. This means that after 24 years, the costs for the PV/storage system are as high as the sum costs over the years of the public grid. Without a battery system, the point is reached after 18 years.

In conclusion, the amortisation of a storage system depends on the PV plant size, the load profile of the household and the electrical storage system itself.

9.5 Conclusions

In this article, simulations to increase SC with a grid-connected PV power plant are illustrated by calculation of an exemplary household. To further increase the level of SC rate and the profitability of the system, the PV system can be combined with a storage system. As a result, the surplus energy of the PV power plant can be stored in the battery and discharged again when energy is needed.

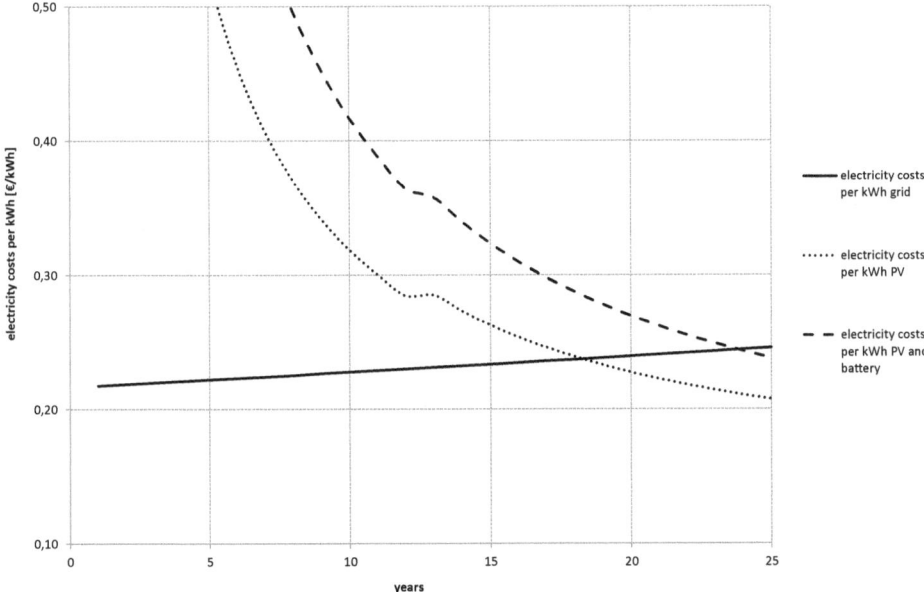

Fig. 9.5 Electricity costs per kilowatt hour

The battery was selected based on the following criteria:

- DoD
- Lifetime
- SC
- Investment costs
- Payback period

There are many new battery technologies under development, and in the next few years, the price per kilowatt hour is expected to drop significantly so that the system can be economically viable within the lifespan of the PV plant.

The mentioned parameters (SC rate and solar-coverage rate) can reach 100 %, but major efforts will have to be made to reach this maximum and it may not be economically attractive. The aim is to find a balance between optimization of the parameters and the reduction of expenses for the system.

The actual behaviour of the user has a major influence on the SC rate. An identical load profile but different annual energy consumption shifts the curve (bigger consumption means higher SC). It turned out that currently an electrical storage system for households only in a few cases is economically viable and energetically appropriate.

Acknowledgments This article is based upon work carried out within the cooperative research project "Vision Step I" (smartcityvillach.at). This project is funded by the Austrian Climate and Energy Fund within the program "SMART ENERGY DEMO – FIT for SET".

References

1. QUASCHNING, V. (2013) Regenerative Energiesysteme—Technologie, Berechnung, Simulation, München: Hanser Verlag.
2. EINFALT, A. et al. (2012) ADRES Concept – Konzeptentwicklung für ADRES – Autonome Dezentrale Regenerative Energie Systeme, Klima- und Energiefonds, Wien.
3. KATHAN, J., STIFTER, M. (2010) Increasing BIPV self-consumption through electrical storage – feasible demand-coverage and dimensioning of the storage system, 5th International Renewable Energy Storage Conference IRES, Berlin.
4. MULDER, G., DE RIDDER, F., SIX, D. (2010) Electricity storage for grid-connected household dwellings with PV panels, Solar Energy 84, 1284–1293.
5. JOSSEN, A., GARCHE, J., SAUER, D.U. (2004) Operation conditions of batteries in PV applications, Solar Energy 76, 759–769.
6. STATISTIK AUSTRIA (2013) Strom- und Gastagebuch: Haushaltsgroßgeräte als bedeutendste Stromverbraucher [Online]. Available from: http://www.statistik.at/web_de/dynamic/statistiken/energie_und_umwelt/071019. [Accessed: 30 September 2013].
7. SAUER, D.U. (2011) Dezentrale Energiespeicherung zur Steigerung des Eigenverbrauchs bei netzgekoppelten PV-Anlagen, Institut für Stromrichtertechnik und Elektrische Antriebe, Aachen.
8. PV-AUSTRIA (2013) Photovoltaic Austria Federal Association [Online]. Available from: http://www.pvaustria.at/. [Accessed: 25 September 2013].

Green Crowdfunding: A Future-Proof Tool to Reach Scale and Deep Renovation?

Sara Kunkel

Abstract

Green crowdfunding is an innovative financing vehicle that allows small investors to contribute towards the improved energy performance of buildings and, at the same time, provides an attractive source of funds for project owners or developers. Social media is of key importance for marketing crowdfunding projects. Some first success stories show how crowdfunding can be an attractive and sustainable business model. Nevertheless, time will tell whether or not this business model can handle more ambitious and comprehensive renovation strategies.

10.1   Introduction

Buildings are central to the EU's sustainable energy policy. They account for 41 % of Europe's final energy consumption and are responsible for 36 % of greenhouse gas emissions. Analysis by Buildings Performance Institute Europe (BPIE) [1] points to the untapped economic opportunities that exist in improving the energy performance of Europe's buildings. Between now and 2050, the average annual investment potential in renovating the existing stock alone could be as high as € 90 billion, not counting the value of high-energy performance in new construction. The International Energy Agency (IEA) and the Joint Research Centre (JRC) have released similar findings.

S. Kunkel (✉)
Buildings Performance Institute Europe (BPIE), Rue de la Science 23, Brussels 1040, Belgium
e-mail: sara.kunkel@bpie.eu

© Springer Fachmedien Wiesbaden 2015
G. Dell, C. Egger (eds.), *World sustainable energy days next 2014,*
DOI 10.1007/978-3-658-04355-1_10

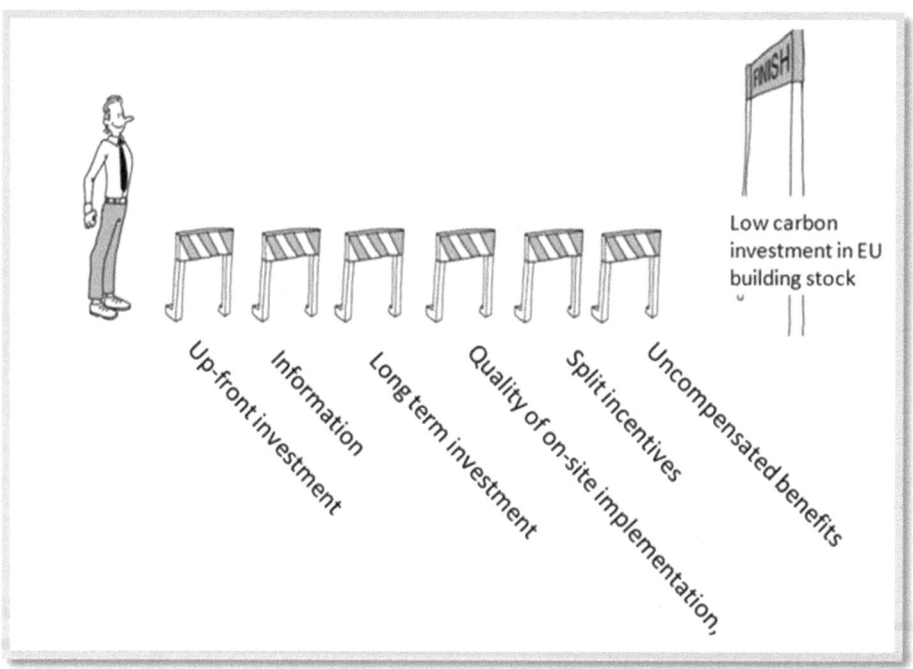

Fig. 10.1 Barriers of low carbon investment in the building stock. (Source: BPIE, Sara Kunkel [2])

There are also other equally important benefits from improving energy efficiency, job creation, health and productivity improvements and alleviation of fuel poverty, to name a few. Yet, its upfront costs act as a significant barrier, particularly for deep renovation. Even though such investments are mostly cost-effective when taken over the lifetime of the measures, they are often stymied by a lack of access to capital (Fig. 10.1).

Large financial institutions are not often interested in small-scale investment, especially when the investments are technically complex, which is the case for Europe's building stock. And while large wind farms receive special rates at banks; a few solar panels on the roof of a school or new lights rarely get a closer look from the banks. Credit is often only available in the pool with other projects or at miserably high interest rates.

Another challenge is that people willing to invest in sustainable retrofitting projects hardly find suitable products on the traditional financial markets. But how can both, money and projects, come together?

Crowdfunding provides a financial vehicle that enables great renewable and energy efficiency projects to find finance outside traditional financial institutions.

Involving the local community in project financing provides new opportunities to drive the energy transition. Crowdfunding can overcome some of the traditional financial and nonfinancial barriers. But is it a future-proof tool to reach the scale of investment we need?

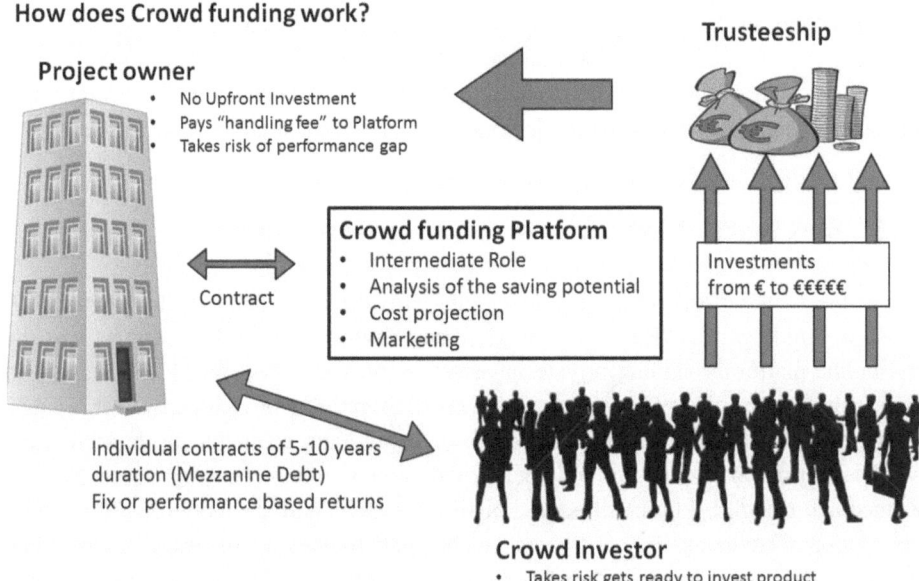

Fig. 10.2 Theoretical scheme of a green crowdfunding business model. (Source: BPIE, Sara Kunkel [2] own illustration)

10.2   What Is Green Crowdfunding and Does It Work?

In general, "Crowd funding is the joint effort of persons who pool their money, typically via the Internet, to support efforts initiated by other people or organizations" [2]. It helps project developers and citizen initiatives find finance sources from their local communities instead of taking a bank loan.

The main advantage of crowdfunding is that, especially in a local context, the investment has positive side effects. Funders become ambassadors of the project they support and they help to promote it through their own networks. According to the European crowdfunding network "the funder usually identifies with the project, has a mind for change and is happy to help provide the social proof of concept".

As shown in Fig. 10.2, the central element of a crowdfunding mechanism is an electronic platform serving as an intermediary tool. Ideally, that is a group of experts from different backgrounds: energy efficiency, finance, law, marketing and social media. The team behind the crowdfunding platform seeks potential projects and gets in contact with building owners. Alternatively, interested building owners get in touch with the platform and propose cooperation. Both project parties have to decide on responsibilities and share of work. Often, crowdfunding platforms facilitate the energy performance improvement measures from planning to implementation. They develop energy concepts based on their own in-house knowledge or outsource the task to external experts. Together with the technical planning, a financing and marketing concept needs to be developed. In the best case, the crowdfunding platform offers a holistic approach based on experiences already made with or without external experts. This way, the building and project owner gets a "one-stop-shop" service and he/she does not have to deal with coordination. In order to cover partly the service of the platform, a handling fee can apply. Once the project framework is set up and contracts are signed, online marketing and the search for investors can start.

10.3 Best Practice Examples of Green Crowdfunding

One of the first start-up companies developing crowdfunding concepts for energy efficiency in buildings was "Bettervest" in Germany. Founded in 2012, Bettervest was the first online platform enabling private investors to put their own money (min. 50 €) in community energy efficiency projects and to profit from the achieved energy cost savings.

The earmarked project money finances ecologically and economically feasible measures delivering cost reduction and energy and carbon emission savings. Measures are planned and calculated by certified and qualified experts and are monitored throughout the duration of contracts. Project owners are obliged to transfer the biggest share of annual savings to the crowd investors, until the upfront investment plus interest rates are paid back. Once the contract ends, energy costs savings stay with the project owner.

The for-profit company Bettervest generates its gains with a finder's fee, a percentage of initial investment costs and generated savings.

Their successful projects already number more than eight. The ranges of implemented measures, initial investment costs and number of investors varies largely from project to project. For all projects, required capital has been collected within a couple of days. Table 10.1 provides a short overview of Bettervest projects and their conditions [3].

As the above table indicates, the biggest partner with cooperation in multiple projects is the company "Bodystreet". Thanks to the great success of the pilot project, three more projects have been implemented. This is reason enough to have a closer look at one example of a branch: the replacement of an old lighting system in Frankfurt:

Table 10.1 Interest rates, duration contracts, carbon savings and total investment volumes per project

Project	Interest rate (APR; %)	Duration of contract (years)	Carbon savings (forecast; tons)	Total investment (€)
Bodystreet 1	7.0	7	3.56	5550
Bodystreet 2	7.35	6	3.96	5250
Zookauf	9	6	25.16	23,350
Social impact lab	10	3	9.62	8400
Bodystreet 3	7	4	8.1	11,950
CHP project Lubeck	8	6	132.2	159,550
CHP Hotel Waldspitze	6.5	3	101.1	61,600

Example: Replacement of old lighting with high-efficiency LED Lighting in "Bodystreet", a Gym in Frankfurt, Germany

Fig. 10.3 Nicholas Pawelke,
bettervest GmbH [3]

Fig. 10.4 Nicholas Pawelke,
bettervest GmbH [3]

During an all-day operation, the gym used annually around 3300 h light, 365 days a year even if the sun is shining. This way lighting caused 83 % of the energy consumption and subsequently a big part of energy costs. After the replacement of the lighting system with high-efficiency LEDs, the share of energy consumption reduced to 37%. In addition, a daylight-dependent automation system maximised the use of natural light. Saving estimations were calculated after an on-site visit by the professional partner "Senergie Consult" and the design of the new lighting installations by "Deutsche Eco Licht".

Costs of the project have been calculated at € 5250 expected energy cost savings are € 1642 each year (based on an energy price of 23.89 ct/kWh). During the first 6 years after the upgrade, 80% of the annual energy cost savings will be paid back, leaving a return of 7.35%. Investments will contribute to a saving of about 6.88 MWh and 3.96 t CO_2 (Figs. 10.3 and 10.4) [3].

10.4   How Crowdfunding Can Overcome Sector-Specific Financial and Nonfinancial Barriers

10.4.1   Barrier 1: Upfront Investment and Bankability of Projects

Relatively high upfront costs are manageable in crowdfunded projects as money can be pooled from a number of investors. Small individual investors get the chance to put their money directly into projects they wish to support. Smaller projects, which traditional financial institutions might consider as not bankable due to high transaction costs, can be financed this way. The best practise example of "Bettervest" shows how a relatively small project can be successfully implemented. Crowdfunding projects and their creditworthiness are not rated according to traditional banking criteria (risk, return on investment). It is the "crowd" that decides which refurbishment project can be realised. As a consequence, an expected return on investment could be significantly lower than for financial institutions. Instead, energy refurbishments, especially when they are local, have an additional emotional value to the crowdfunder. With the service provided by crowdfunding platforms the process of investing in energy efficiency projects is simplified as projects are usually "ready to invest".

10.4.2   Barrier 2: Lack of Information

Crowdfunding helps sustainable projects find finance from individuals by providing them with good information on the project in a comprehensive and objective manner. By making optimal use of online marketing channels (including YouTube and Facebook), project developers are able to update investors on latest developments in building and maintaining the projects. In addition, word-of-mouth marketing works well for local energy refurbishment projects. Social media plays a key role for the project's dissemination. At the same time, a project owner can use the publicity as a marketing tool for their businesses.

10.4.3   Barrier 3: Quality of On-Site Implementation and Trust in the Project and Local Companies

Trust in the actual delivery of an energy-saving measure is a big issue for institutional investors. Green crowdfunders usually trust calculated saving and the proper implementation of the energy-saving measure as they might know the project personally or even the company in charge of implementing the measures. Depending on the type of contract, investors can either (partly) take the performance risk or go for a guaranteed interest rate when the performance risk is covered by the project owner.

So far, only limited examples of performance-based interest rates exist.

10.4.4   Barrier 4: Split Incentives and Uncompensated Benefits

Local crowdfunding leads to greater local growth. Increased local economic activity results in higher income from duties. Where the public sector is involved in energy-saving measures, budgets can be saved and reassigned.

This is especially the case for renewable energy projects that in some countries are subsidised with feed-in tariffs, which citizens have to pay via their electricity bills (e.g. Croatia, Germany); it is important to get local people on board. In countries where citizens have very little ownership over the renewable power capacity being built, most of the governmental incentives flow out of the country. In effect, the citizens are financing the renewable energy boom, without enjoy direct benefits.

10.5   Think Global, Act Local? What We Can Learn from the Crowdfunding Examples

Green crowdfunding is a promising model because it gives individuals a new place to directly invest in sustainable projects that are real and create lasting value. It is a unique opportunity for small investors, especially for investors without their own property, to contribute to an improved energy performance of European's building stock. The amount of people-powered finance which has been raised for example in Germany shows that it is also possible for other countries to make a major and direct contribution to the capital investments we need for a low-carbon energy transition.

As crowdfunding platforms deliver "ready-to-invest" projects, their biggest achievement is bringing different market actors together. First success stories exist, but the business model has to stand the test of time. So far, all pilot projects focus on "low-hanging fruits", meaning measures with a relatively short payback period, such as the replacement of lightning. To tackle the full challenge of Europe's buildings' stock, it seems to be desirable to extend future projects on comprehensive refurbishment. Deep renovation so far has not been realised with crowdfunding.

For the future of crowdfunding, more routine and standardisation is needed to reduce planning time and transaction costs. Start-up companies with first experiences have been able to reduce their planning time project by project. First success stories created trust and made it easier to forecast savings that can be expected. Nevertheless, more data are still needed and it is highly recommended to include long-term monitoring in project creation.

Decision makers need to understand that energy efficiency is not a burden, but an investment opportunity with a great rate of return.

References

1. ECONOMIDOU, M. et al. (2011) *Europe's Buildings under the microscope.* Brussels. Buildings Performance Institute Europe.
2. KUNKEL, S. (2014) Buildings Performance Institute Europe, Brussels.
3. EUROPEAN CROWD FUNDING NETWORK (2012) [Online] http://www.europecrowdfunding.org [Accessed: 19 March 2014].
4. MIJNALS, P. (2013) *Bettervest,* [Online] http://bettervest.de/de/content/projekte/ [Accessed: 10 June 2014].

Part II
World Sustainable Energy Days Next Conference 2014: Biomass

Satu Lantiainen and Nianfu Song

Abstract

The comparative analysis shows that both the EU and the USA have a wide range of energy policies that allow flexibility in how targets are met. While the USA has renewable electric power and biofuel targets, EU has targets for renewable energy including heating and electric power and biofuel. The USA uses policy instruments to promote mainly electricity and biofuel. Contrary to policies in the USA, biomass energy policies in the EU emphasize all types of renewable energy output including heating.

11.1 Introduction

Woody biomass is the largest source of renewable energy in the European Union (EU) and the USA. In 2011, wood and wood wastes contributed 49% of all renewable sources in the gross final energy consumption in the EU-27 [1]. In the USA, the wood energy share of renewables reached 22% in 2011 [2]. However, in the EU and the USA, the relative share of wood energy in total renewable energy sources has decreased over the years—in the EU-27 from 56% in 1990 to 49% in 2010 as the use of other renewable energy sources has grown at a faster rate [1]. In the USA, the dramatic increase of use of corn ethanol as a transportation fuel decreased the relative share of wood energy in total renewable energy sources. Concern over greenhouse gas (GHG) emissions and climate change, energy secu-

S. Lantiainen (✉)
University of Missouri, 239 Anheuser-Busch Natural Resources Bldg., Columbia, MO 65211, USA
e-mail: lantiainens@missouri.edu

N. Song
University of Missouri, 237 Anheuser-Busch Natural Resources Bldg., Columbia, MO 65211, USA
e-mail: songn@missouri.edu

© Springer Fachmedien Wiesbaden 2015
G. Dell, C. Egger (eds.), *World sustainable energy days next 2014,*
DOI 10.1007/978-3-658-04355-1_11

rity, and lack of energy supply diversity has motivated the development of public policies promoting the use of woody biomass in energy generation in both regions.

The EU and the USA have set targets for renewable energy use. In the EU, the directive on the promotion of the use of energy from renewable sources (Renewable Energy Strategy (RES) directive) from 2009 sets binding targets for renewable energy production for all member states. In the USA, rather than a federal mandate, individual states have adopted mandatory and voluntary renewable energy targets or none at all.

Most recent studies evaluating EU energy policies have focused on general renewable energy policy instruments on heating, electricity, and transport. Kitzing et al. [3] reviewed national support policies for electricity from renewable energy sources in the EU and found out that policies show a significant tendency of convergence. Connor et al. [4] have analyzed different kinds of support options for renewable energy sources of heating in the EU and suggest that wider regulations must be considered. In the USA, the current policy instruments promoting wood-to-energy use have been examined and analyzed. Aguilar and Saunders [5] suggested that that among policy instruments adopted by individual states, regulatory policies have relied primarily on the adoption of renewable portfolio standards (RPSs) mandating the generation of energy from nonfossil fuels. Aguilar et al. [6] point that at the US federal-level financial incentives have been the most common tool to promote greater wood energy generation.

The main objective of this study was to compile, review, and compare the multiple public policy instruments promoting the sustainable use of woody biomass for energy generation in the EU and USA. In the EU, special attention is paid to select case countries: Finland, Germany, and the UK.

11.2   Wood Energy Public Policy Review

11.2.1   Wood Energy Public Policies in the EU

The EU energy and climate package was adopted in 2008. The energy and climate package sets four main targets for the EU member states by 2020: 20% of energy to come from renewable sources, 20% reduction in EU GHG emissions from 1990 levels, 20% improvement in EU's energy efficiency, and 10% of transport fuels from renewable energy. In order for the EU as a whole to reach 20% of energy from renewable sources, member states have taken on legally binding national targets. These targets range between 10 and 49% depending on a starting renewable energy production and consumption point of each member state. Despite the renewable energy policy at the EU level, member states have the competence on deciding upon their own national policies to boost the renewable energy consumption and set even more ambitious renewable energy targets. Feed-in tariffs (FITs) have become a dominant policy mechanism in the renewable electricity sector in the EU. To date, 23 countries in the EU-27 have a FIT scheme in place. Heat premiums are less common but are becoming more popular.

There are deductions, exemptions, and reduced tax rates available for renewable sector in the EU member states. The range is wide, varying from tax deduction on labor costs and equipment to exemptions from fossil fuel taxes. Germany offers tax reliefs for certain energy products in efficient combined heat and power (CHP) plants. In Finland, renewables are exempt from energy tax on heating sector, whereas the UK has an environmental tax for users of nonrenewable electricity. All three case countries have incentive programs for residential users of woody biomass heating. Direct input subsidy for woody biomass is a rare policy tool in the EU. Of the case countries, Finland has outlined a new legislation on input subsidy for power plants.

The Common Agricultural Policy (CAP) of the EU has specific forestry and bioenergy measures. In the current CAP period, 2014–2020, subsidies on both the national and the EU level could have been allocated to measures such as installations/infrastructure for renewable energy using biomass and processing of forest biomass to renewable energy.

11.2.2 Wood Energy Public Policies in the USA

The foundation for wood and renewable energy regulations consists of several regulatory policies, such as the Energy Policy Act 1992 (EPACT1992, Pub. L. No. 102-486), the Energy Policy Act 2005 (EPACT2005, Pub. L. No. 109-58), and the Energy Independence and Security Act of 2007 (EISA 2007, Pub. L. No. 110-140). Renewable standards, public benefit funds, renewable power metering, electric power interconnection and construction regulations, and mandate for transportation fuels are some of the main regulations in the state level. For example, to date, 30 states have enacted mandatory RPS or other mandatory renewable energy capacity policies.

US financial incentives are provided to both consumers and producers. Fourteen out of 18 of the US federal renewable energy incentives, 210 out 445 state incentives, 2 out of 3 federal renewable regulations, and 189 out of 388 state renewable regulations are applicable to biomass energy including woody biomass energy. Grants in the USA are disbursed often from funds of federal, state, or local government agencies to biomass energy projects. By June 2012, there were four federal grants applicable for woody biomass power or CHP projects. Twenty-five US states also offer biomass energy grants for biomass energy.

11.3 Comparative Analysis of Wood Energy Policies

The USA and the EU have financial incentives and preferential regulations for renewable energy including biomass energy. Renewable standards or targets, FITs, subsidies, tax exemptions, and low-interest loans are common to these countries. State-, provincial-, county-, or city-level incentives and regulations exist along national/EU-level biomass energy policies. The USA uses policy instruments to promote mainly electricity and biofuel. Contrary to policies in the USA, biomass energy policies of the EU member states emphasize all types of renewable energy outputs including heating. While the USA only has renewable electric power and biofuel targets, EU has targets for renewable energy including heating and electric power and biofuel.

Both the USA and the EU bundle woody biomass policies with other biomass feedstock, such as agricultural feedstock. Policies that promote particular technology and biomass type can distort the markets [5]. Also, policies are meant to be applicable in different kind of areas, despite of the type of the feedstock being produced. However, EU case countries seem to have some particular programs, e.g., UK's miscanthus incentive.

Tax instruments are used in both the USA and the EU member states (the responsibility of direct taxation lies within the EU member states). The tax credit and grants in USA are similar to subsidies in the EU member states. For example, the production subsidy in Finland is similar to the production tax credit in the USA, but unlike the USA whose production tax credit available only to electricity production, Finland production subsidies provide support for both heat and electric power production. Similar to the investment credit in the USA, investment subsidies are also available in the EU member states and also promote both heating and electric power generation.

Low-interest loans and tax exemptions are commonly used in the USA and EU member states. US loans are mainly used in biomass electric power, but EU loans are used for both heating and biomass electric power.

Major US incentives and regulations such as federal green power purchasing mandate required by FIT, state RPS, bonds, loans, grants, and net metering regulations are helping mainly renewable electric power including woody biomass electric power. CHP uses residual heat and therefore has higher thermal efficiency. However, pollution concern and huge investment on the heat distribution system make a new CHP facility close to community very unlikely in most cities. The pulp and paper industry is now taking the advantage of the CHP preferential policy and using residues derived from manufacturing as feedstock for CHP [7]. Even so, the capacity of inefficient biomass electric power has grown faster than high efficient CHP biomass power in 1980s, and the proportion of CHP capacity has changed between 34 and 38 % since 2000, and the current proportion of CHP is 34 %, the lowest since 2000 [7, 8].

The European incentives, such as those of Finland and the UK, for households switching from oil or electricity heating to wood heating are examples of promotion for both biomass energy and efficiency. Compared to the increasing woody biomass consumption in the EU, the declined US consumption reflects the differences in supporting heating biomass technologies.

In the USA, many incentives are available for energy generation from various energy sources, not necessarily only for renewable energy generation. For example, the existing Residential Energy Efficiency Tax Credit gives no advantage to biomass stoves over fossil fuel stoves in the residential sector, and thus does not contribute to replacing fossil fuels. Moreover, unlike the USA, the EU addresses, or plans to address, sustainability concerns in the renewable energy policies.

The main target of renewable energy policies is to increase the energy generation from renewable energy sources. Figure 11.1 shows a continuous increase of woody biomass en-

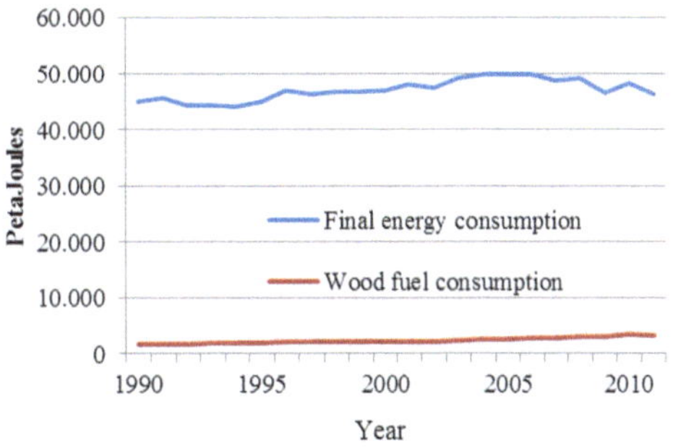

Fig. 11.1 Final energy consumption and wood fuel consumption in EU-27, 1990–2011. (Source: EURO-STAT [1])

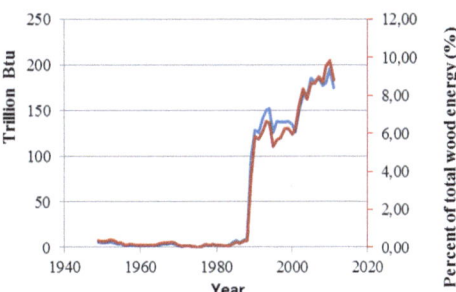

Fig. 11.2 Woody biomass consumption by the US electric power sector from 1949 to 2000, estimated by US EIA [2]

ergy consumption in the EU-27 since 1991. Binding EU policies have not yet contributed to this development and therefore this increasing trend can only be explained by the effectiveness of the national woody biomass energy policies. Recent statistics [1] show that, despite the economic downturn, the use of renewable sources in energy generation has increased during 2009 and 2010, after the introduction of the binding renewable energy targets in the EU.

The incentives and regulations in the US states in the 1980s targeted mainly renewable electricity generation, and most of the biomass power facilities enjoyed the mandate electricity purchase by utilities entities. The result was a jump in the number of biomass power plants and biomass consumption by commercial and electricity utility energy consumption sectors in the USA. The historical increase of wood energy consumption in the late 1980s and 1990s, as shown in Fig. 11.2, was mainly a result of state incentives and regulations newly established Public Utility Regulatory Policies Act of 1978 (PURPA).

The wood energy consumption by electric power started to go up again after 2002 when energy price was high and ten additional states (e.g., California, Florida, Maine, Michigan, and Texas) enacted RPS 5 years before 2002. While the increasing trend in wood energy consumption by electric power reflected the effects of biomass policy efforts, the declining trend of the residential wood energy consumption in the late 1980s and 1990s, and the downtrend of wood energy use by the US industry after 1990 dominated the total woody biomass consumption in the USA. Because majority of woody biomass was used by the industrial and residential sectors that public incentives and regulations do not focus on, the impact of renewable policies on wood energy consumption is still insignificant until now.

11.4 Conclusions

Both the EU and the USA have a set of renewable energy policies that are aiming at increasing the share of renewable energy sources, including woody biomass, in energy generation. The main regulatory policies, RES directive in the EU and RPS in the USA allow flexibility in how targets are met.

The analysis shows that there are also energy policies that hinder the effectiveness of existing renewable energy policies. Public support to low-efficient energy conversion or use of feedstock that only provide marginal GHG net reductions compete for limited public resources that could otherwise be used to further utilization of highly effective biomass technologies. In addition, changing market conditions directs the use of energy resources. Growing imports of fossil fuels (e.g., coal) due to lower international prices and the collapse of the EU emission trading system risk the reduction of GHG emissions from the EU. Therefore, policies favoring the use of renewables, including woody biomass, need to be further developed and the complex energy policy landscape comprehensively reassessed.

References

1. EUROSTAT. (2012) *Renewable energy—an analysis of the latest data on energy from renewable sources.* [Online] Available from http://www.epp.eurostat.ec.europa.eu/cache/ITY_OFFPUB/KS-SF-12-044/EN/KS-SF-12-044-EN.PDF [Accessed 30 November 2012].
2. US EIA. (2013) U.S. Energy Information Administration (US EIA) (2013) 'Feed-in tariff: a policy tool encouraging deployment of renewable electricity technologies'. [Online] Available from www.eia.gov/todayinenergy/detail.cfm?id=11471 [Accessed 22 August 2013].
3. KITZING, L., MITCHELL, C., MORTHORST, P.E. (2012) Renewable energy policies in Europe: converging or diverging. *Energy Policy*. 51. 192–201.
4. CONNOR, P. et al. (2012) Devising renewable heat policy: overview of support options. *Energy Policy*. 59. 13–16.
5. AGUILAR, F.X., SAUNDERS, A. (2010) Policy instruments promoting wood-to-energy uses in the continental United States. *Journal of Forestry*. 108. 132–140.
6. AGUILAR, F.X., SONG, N., SHIFLEY, S.R. (2011) Review of consumption trends and public policies promoting woody biomass as an energy feedstock in the U.S. *Biomass and Bioenergy*. 35. 3708–3718.
7. OAK RIDGE NATIONAL LABORATORY (ORNL) (2012) Current Biomass Power Plants. [Online] Available from cta.ornl.gov/bedb/biopower/Current_Biomass_Power_Plants.xls [Accessed 1 November 2012].
8. BOUNDY, B. et al. (2011) *Biomass Energy Data Book. 4th Ed.* Oak Ridge National Laboratory. Oak Ridge, Tennessee.

Biomass Opportunities and Potential in Northern British Columbia, Canada

P. Sean Carlson

Abstract

The use of biomass as a heating fuel in northern British Columbia (BC) is an attractive alternative to traditional fossil fuels. The benefits of using biomass in BC include cost savings of at least 50 %, reduced greenhouse gas emissions, and increased energy security in individual communities and opportunities for local employment and engagement. Using the Austrian biomass industry as a model to establish and grow the industry in northern BC, many of the early issues can be avoided, such as over-dimension, resulting in a highly productive and robust network. Achieving a high level of growth in the biomass industry in northern BC would require the involvement of the various levels of government, private businesses, and educational institutions and most importantly, the citizens of the communities it serves. Financing, technical expertise, fuel sources, and biomass technology are all available—it is a matter of engaging local municipalities and First Nation communities by promoting biomass heating as an economic and sustainable alternative

12.1 Introduction

The biomass energy industry in northern British Columbia (BC) is in its early stages of development and requires a clear strategy for growth. Using the development and growth of the Austrian biomass energy industry over the past 30 years as a model economy, it

P. S. Carlson (✉)
University of Northern British Columbia, 3333 University Way,
V2N 4Z9 Prince George, BC, Canada
e-mail: patrick.sean.carlson@gmail.com

© Springer Fachmedien Wiesbaden 2015
G. Dell, C. Egger (eds.), *World sustainable energy days next 2014,*
DOI 10.1007/978-3-658-04355-1_12

will form the basis of a series of recommendations for the development of the industry in northern BC. The goal of the report will be to demonstrate how the growth trend currently experienced in Austria can be achieved throughout northern BC.

12.2   Biomass Energy

12.2.1   Need for Biomass

The economic drivers of northern British Columbia (BC) are reliant on both the finite and recurring resources oft the natural environment. Fishing, forestry, mining and tourism are industries that rely, in some capacity on either renewable or non-renewable resources. The lifestyle of the First Nation people demonstrates the stewardship of the natural resources available from the land and sea that support their way-of-life. This way-of-life is comparable to the 21st century definition of sustainability. However, the region is challenged to find a balance—the extraction and sale of finite resources with the preservation of the natural surroundings and social fabric of the communities. According to the Invest in Northwest BC website, there is US $ 60 billion of investment proposed or planned for the region. Balancing the social and environmental benefits of our natural surroundings with the current rate of the resource development in northern BC for both domestic and international energy demands requires that sustainable alternatives be industrialized.

One solution includes limiting or completely eliminating our use of finite fossil fuels as our primary energy source by transitioning to different types of available renewable energy. Solar, wind, tidal, and biomass energy are underutilized or underdeveloped in northern BC. If we sustainably manage our renewable energy sources, we could balance the economic needs of the province with the social and environmental characteristics of our natural surroundings.

In 2011, Natural Resources Canada (NRC) and Aboriginal Affairs and Northern Development Canada (AANDC) produced a report summarizing the current status of remote and off-grid (those not connected to the electricity grid) communities in Canada [8]. In BC, there are 86 sites that are off-grid, with an average electricity price of US$ 0.37/kWh. For on-grid customers, the current residential electricity rate charged by BC Hydro for the initial 1350 kWh of power consumption over a 2-month billing cycle is US $ 0.069/kWh. Off-grid communities, such as Masset, BC, are paying approximately five times more for power compared to Prince George. Although the NRC and AANDC reports refer to the distribution of electricity, the same challenges apply when trying to supply communities with heating fuels that do not have access to a natural gas distribution network. Fuel costs, energy security, environmental impacts, health benefits, and local employment are all issues that could be addressed when planning alternative energy projects.

12.2.2   Results of the Biomass Industry

Replacing fossil fuels with locally sourced biomass could positively contribute to the northern BC economy. First, it could provide local residents with value-added logging and processing jobs. New types of jobs could be created in the communities to support the biomass industry. Second, obtaining heating fuels from local sources could result in increased energy security. With local sourcing of biomass, communities would be less susceptible to the global increase in fossil fuel prices. Third, by reducing the cost of heating fuels and generating local jobs, more money would stay in communities. Lastly, by utilizing locally available, renewable fuels, communities will save on the carbon tax at US $ 30/t of CO_2 equivalent.

The intent of developing a biomass industry is to provide communities with the opportunity to eliminate their dependence on outside sources of energy, create local jobs for residents, manage resources sustainably, and provide better services through energy cost savings [9]. Successfully transitioning from fossil fuels to renewable biomass may need to be promoted and financially supported in communities with local access to biomass, such as Prince Rupert, Port Edward, Lax Kw'alaams, Metlakatla, Kitsumkalum, Terrace, the "Hazeltons," Gitanyow, Smithers, and Fraser Lake.

12.3   Bioenergy in Austria

Focusing on the history of the development and ongoing trends of the biomass industry in Austria and describing the current status of the industry in northern BC allows for a series of recommendations regarding how to expand the biomass industry in northern BC concurrently with the Austrian industry. Although the Austrian situation will not be identical to that in BC, similar problems will be encountered. The objective is to model the growth of the biomass industry in BC off of the growth trend experienced in Austria.

12.3.1   Growth

From Table 12.1, Austria and BC have comparable heating degree days; however, BC has a much greater volume of forested area which could be sustainably managed in part for bioenergy production. In 2009, Austria's overall consumption of energy and nonenergetic resources totaled 41 million t of oil equivalent (Mtoe), of which 29 Mtoe is imported [2]. Of the total consumption, 29 % is from renewable sources with the remaining 71 % coming from fossil fuel sources, making Austria's economy vulnerable to fluctuations in the price of fossil fuels or disruptions in supply. The government in Austria is trying to develop local sources of energy, specifically renewable sources such as solar, wind, and biomass to achieve energy independence.

Table 12.1 Comparison of Austrian and British Columbia (BC) trends related to biomass energy [1, 5, 8, 7, 11]

	Austria	British Columbia
Population	8.6 million	4.4 million
Land area	84, 000 km²	945,000 km²
Heating degree days (>18°C)	3500	4000
Forest area	40,000 km²	550,000 km²
Agriculture area	30,000 km²	47,000 km²
Annual bioenergy production	5.9 Mtoe	3.7 Mtoe

In the 1980s, Austria was heavily reliant on imported fossil fuels, such as oil. By 2011, domestic bioenergy production accounts for approximately 14.0% of the total energy demand. From 1980 to 2011, the production of bioenergy in Austria grew from 1.1 to 5.9 Mtoe, an increase of more than 400% over 31 years.

An important milestone for the bioenergy industry in Austria occurred when state governments recognized the importance of the industry and contributed to it financially [3]. Without governmental support, the bioenergy industry would not have been able to maintain a stable market to attract potential customers. In addition, both the state and federal governments contributed financially to the industry through grants and loan repayment programs [6]. These programs allowed private entrepreneurs to become designers, builders, and owners of biomass district heating (BDH) plants.

The success of the Austrian bioenergy industry has allowed different levels of the Austrian government to provide funding to expand the industry internationally through events such as seminars, courses, and conferences. In addition, Austrian companies recognize that to be successful, there needs to be a partnership between the communities, universities, and governments in target markets.

12.3.2 Early Issues

In the early years of development, many of the project planners and engineers were lacking in appropriate experience to design biomass heating plants:

> As a consequence, in the early years many planning and installation mistakes were made, the most common probably being over-dimensioning of the systems, sometimes leading to excessive project costs and poor economic performance. [6]

During the initial years, most planners and engineers did not understand the potential issues resulting from oversizing biomass boilers. Not only does over-dimensioning of biomass boilers lead to poor economic performance and excessive project costs as stated by Madlener, it can also lead to increased emissions output when boilers are not operating in the optimal range [4, 6]. Engineers failed to accurately determine the heating demand of a facility and did not account for the 4:1 turndown ratio of most biomass boilers. Accessing heating data, such as past utility bills for natural gas and electricity consumption, and climate data for the area in question will help engineers accurately size a biomass boiler system.

Sustainable harvesting of timber is an issue in Austria that has yet to be resolved. Government-owned forests are being harvested at an unsustainable rate. If biomass is overharvested, the industry will face similar challenges as before—the need for imported fuels.

12.4 Bioenergy in Northern BC

12.4.1 Areas of Potential Growth

Three areas of potential growth for biomass heating systems (BHS) in northern BC are: towns and cities without access to a natural gas distribution network; homeowners and businesses replacing old, inefficient boilers; and business opportunities building new district heating systems. Optimizing the growth of the biomass industry requires the projects with the greatest economic benefits to be identified and completed first. Therefore, the priority could be to implement BHS in towns and cities without a natural gas distribution network. As the industry establishes itself financially and the cost to design and build a BHS system decreases, it will open up opportunities in areas that were previously cost prohibitive, such as in Prince Rupert and Terrace.

Biomass-powered district heating systems could represent a method to grow the biomass industry. District heating is a system of providing heat through a network of pipes. One of the current obstacles to implementing district heating systems in BC is the uncertainty of costs for construction of the piping network. In the majority of feasibility studies, the cost per meter to install piping is the limiting factor. It could be worthwhile to further investigate this constraint. If these systems become more common and the idea of selling heat, rather than a heating fuel, becomes familiar to the general population, it will be easier to promote district heating systems in areas where it would not otherwise be economic.

12.4.2 Economics

The cost of heating fuel in northern BC is the primary driving force behind the growth of the biomass heating industry (see Table 12.2). In communities that need to have heating

Table 12.2 Cost of selected heating fuels/energy in northern BC [10]

		Natural gas ($)	Propane ($)	Electricity ($)	Pellets ($)	Wood chips ($)
Subtotal	/GJ	14.19	20.00–40.00	22.20	8.00	3.03
Carbon tax	/GJ	1.564	1.564	0.00	0.00	0.00

fuels delivered by truck rather than distributed via a larger network, the cost can be greater than US$ 40/GJ which is over twice the cost of natural gas from the Prince Rupert to Vanderhoof Pacific Northern Gas (PNG) Ltd. pipeline [10]. The majority of these communities have easy and inexpensive access to biomass, making the cost savings even greater. By developing fuel supply chains in each community, similar to the cooperative system in Austria with the "biomass trade centers," it will provide local employment which is needed in some northern BC cities.

The cost savings between the current prices of heating fuels and the prices of biomass is what makes BHS worthwhile. The cost to implement a BHS is higher than a similar-sized natural gas or electric system due to the robustness of the technology required. However, these higher initial costs are offset by the fuel savings and the less tangible benefits of creating a sense of ownership in the local economy through value-added jobs.

12.4.3 "Biomass Deployment" in Northern BC

To be successful in northern BC, the biomass industry will need support from the communities; local, regional, provincial, and FN governments; universities and colleges; the forestry industry; and current fuel suppliers. The universities and colleges could train the individuals to design, build, and operate the BHS plants and act as a "model" community by implementing biomass heating facilities to demonstrate how to properly size, build, and operate a BHS plant. Various levels of government could provide the financial and regulatory support to ensure these projects are implemented in a timely and efficient manner. The forestry industry could benefit from the development of an additional market to sell what is currently a waste product. Current fuel suppliers can act as investors in BDH projects, helping to diversify their business portfolio. Ultimately, none of these projects will be successful without the support of the local community. They must be convinced of the social, economic and environmental benefits of the biomass industry in order for it to spread across northern BC. Table 12.3 provides a summary of the current state of the bioenergy business in BC as compared to Austria.

Table 12.3 Qualitative comparison of the supporting element of the biomass industries in BC and Austria

	Austria	British Columbia
Governmental policy	Well established	In development
Trained personnel	Well established	In development
QA/QC program	Developing	Not available
Fuel supply network	Well established	In development
Boiler manufacturers	Many	Few
Consumers	Many	Many

QA/QC quality assurance/quality control

Madlener highlighted nine points to promote the diffusion of BHS in the Vorarlberg province in Austria that can be applied in northern BC:

1. Local promoters of BDH projects
2. The agent(s) acting in each province as a focal point
3. Skilled planners and installers
4. Capital grants for BDH systems with a community component
5. Capital grants for connecting residential homes to BDH systems
6. Grants for pre-feasibility studies
7. Environmental protection
8. Convenience
9. Sustainable local development

12.5   Conclusion

There is a need to investigate renewable sources of energy to displace the use of fossil fuels. The three crucial milestones to growing the biomass industry in northern BC could include: building biomass energy systems in communities that lack access to a natural gas distribution network and have an abundant supply of biomass; using well-developed technology and design techniques; and establishing adequate financial support network.

The first milestone could be achieved through public consultation and government backing. The second milestone could be reached through building partnerships between Canadian and Austrian universities and engineering consulting companies. The final milestone could be realized through a combination of government grants and loan repayment programs, private lenders (such as credit unions), and community funds. Forming a network of resources for the production, distribution, and utilization biomass energy in northern BC will be an important component on which to build the enormity of its potential. With the role BC will play in the global energy theater over the coming decades, it is important we promote lasting and sustainable energy choices.

References

1. AUSTRIA: ARRIVE AND REVIVE (2013) *Government and People*. [Online] Available from: http://www.austria.info/us/about-austria/government-people-1140686.html [Accessed 9 September 2013].
2. AUSTRIAN ENERGY AGENCY (2012) *Reports & brochures: Austrian Energy Agency*. [Online] Available from: http://en.energyagency.at/facts-services/publications/reports-brochures.html?no_cache=1 [Accessed September 2013].
3. EGGER, C. et al. (2013) *Biomass heating in Upper Austria. Green energy, green jobs*, Linz, Austria: O.Ö. Energiesparverband.

4. ENVIRON (2009) *British Columbia Air Quality and Health Benefits Report*, Victoria: Province of British Columbia.
5. HANAK-HAMMERL, D. (1996) *FAO Corporate Document Repository*. [Online] Available from: http://www.fao.org/docrep/w3722e/w3722e05.htm [Accessed 1 October 2013].
6. MADLENER, R. (2006) Innovation diffusion, public policy, and local initiatives: The case of wood-fuelled district heating systems in Austria. *Energy Policy*, 1992–2008.
7. MINISTRY OF FORESTS, MINES AND LANDS (2010) *The State of British Columbia's Forests*, 3rd ed., Victoria: Province of British Columbia.
8. NATURAL RESOURCES CANADA (2011) *Status of Remote/Off-Grid Communities in Canada*, s.l.: Natural Resources Canada.
9. PACIFIC INSTITUTE FOR CLIMATE SOLUTIONS (2013) *Fire in the Woods or Fire in the Boiler?* Victoria: s.n.
10. PACIFIC NORTHERN GAS Ltd. (2013) *Rates: Vanderhoof to Prince Rupert/Kitimat.* [Online] Available from: http://www.png.ca/vanderhoof-prince-rupert-kitimat/ [Accessed 23 September 2013].
11. STATISTICS CANADA (n.d.). *Population and dwelling counts, for Canada, provinces and territories, 2011 and 2006 censuses*. [Online] Available from: *http://www12.statcan.ca/census-recensement/2011/dp-pd/hlt-fst/pd-pl/Table-Tableau.cfm?LANG=Eng&T=101&S=50&O=A* [Accessed 9 September 2013].

Eco-Energy Aspects of Production and Utilization of Agripellets

Viktória Papp

Abstract

In addition to the environmental reasons, growing energy demand, running out of fossil fuel supplies, and the expected increase in gas prices, all indicate that we must change in our power supply. We should be increase the renewable energies 14.65 %, up to 2020. The opportunity among the renewable energies available in Hungary largely lies in the utilization of biomass. This compressed energy has come into the purview of Europe and our country too. The EU market is ideal for wood pellet production. In Hungary, due to the characteristics of its agricultural industry, large amounts of herbaceous biomass are available. Straw and various agricultural byproducts can be used as the raw materials for agripellets. Common complications of the various by-products used in pellet production are the ability to store and manage them, in addition to their combustion. Therefore, it is important to create pellets that will reduce the energy put into transportation and improve the combustion parameters. Despite the fact that we have the herbaceous raw material base, the agripellet production is only slowly developing in our country. One reason for this is that while we have various agripellet combustion furnaces and boilers, these systems are relatively expensive. In addition, due to the high ash content of herbaceous plants, it cannot be burned in wood pellet boilers. Furthermore, furnaces in the market are relatively few. However, Austria and a number of EU countries are helping with subsidies for the initial investment to make the changeover to pellet heating. In our studies, we dealt with the biodiesel production generated from the by-product of rapeseed stalk. After the grinding process, we produced rapeseed stalk pellets with a small pellet-making machine. Studies show that we can obtain a lot

V. Papp (✉)
University of West-Hungary, Bajcsy st 4, Sopron 9400, Hungary
e-mail: pocokvick@gmail.com, e-mail: pappviktoria@emk.nyme.hu

© Springer Fachmedien Wiesbaden 2015
G. Dell, C. Egger (eds.), *World sustainable energy days next 2014*,
DOI 10.1007/978-3-658-04355-1_13

of energy from the rapeseed stalk. The location of the examination took place in T&T Technik Ltd. in Szentes. They are producing agripellets from different agricultural by-products. In the future, I would like to expand on energy balance research in the area of agripellet production.

13.1   Introduction

Due to the climate and geological features (characteristics) of Hungary, there are large amounts of wood and herbaceous biomass in which a significant part can be used for energy purposes. The pellet production in the biomass sector is one potential area for solid biomass. The pellets are produced at high pressure, and are compressed into a cylindrical shape. It is characterized by high density and compactness. The diameter of the pellets is 5–10 mm, and they are 10–25 mm in length [1]. During timber processing and furniture production, wood chips and shavings are produced, which can be processed into wood pellets and briquettes. The wood-based resources that are needed to increase the production are limited. However, the agricultural by-products that are generated each year on arable lands are increasing. This is mostly due to the reduction of livestock that would normally use these by-products.

In addition, due to the nutrient content, a part of the by-products, in particular straw and crop residues are returned to the soil after chopping. Nevertheless, lignoceluloses cannot be returned in unlimited quantity into the soil because the excessive recirculation of herbaceous residues causes the penthouse effect. Therefore, the cellulose-decomposing bacteria reduces the nitrogen content of the soil, which allows significant amounts of fertilizer input to be balanced [2]. Thus, the raw material base of agripellet production is given.

13.2   Situation of Pellet Plants in Hungary

The European pellet sector began to develop rapidly in recent years. Between 2000 and 2010, the pellet consumption has increased more than tenfold in the European Union. The number of European pellet plants has reached 700 in 2010. Due to the dynamic development of pellets, it has become necessary to enlarge the base of raw materials used because the quantity of by-products from the wood industry is limited [3]. This is why agripellets have appeared in market.

Agricultural plant materials, residues, and waste are excellent heating fuels in pellet form. These plants are reproduced annually and can be harvested. The ash content of agripellets (made from herbaceous plants) is between 3 and 10%; in addition, the heat values are more varied than wood pellets [4].

Figure 13.1 shows the operating agripellet and wood pellet plants in Hungary. The agripellet production was less than 8000 t in 2010 and this was mostly sold domestically, which is in contrast to the wood pellets that are mostly exported to foreign markets. Unfor-

Fig. 13.1 Wood pellet and agripellet plants in Hungary. (Source: Hungarian Pellet Association [5])

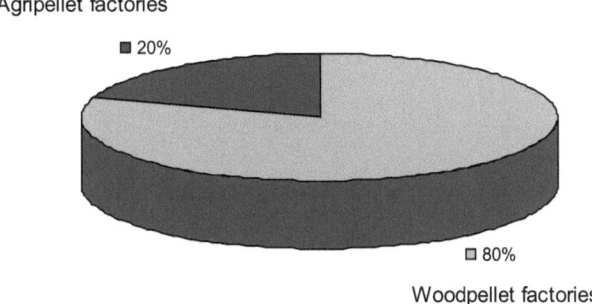

Agripellet factories

■ 20%

□ 80%

Woodpellet factories

tunately, customers often meet with poor quality or friable agripellets. This has worsened the assessment of pellets, which are made from herbaceous plants. The spread of agripellets could help with the initial support of furnace investment. In our country, we should not calculate the availability of raw materials necessary for the production of agripellets, as large amounts of agricultural by-products are available. At the same time, investments in technology and development are necessary for the advancement of the sector. It is necessary to establish a pillar in the "green economy". The companies have already entered the market but the barriers in combustion technology impede the spread of residential utilization. Currently, the market for traditional heating methods is more popular [6].

13.3   Production of Pellets

The pellets are compressed energy that produces energy at a high (800–900 bar) pressure. Depending on the moisture of the raw material, it is common to have to apply some kind of drying technology, which often requires a lot of energy. A specific technology developed in Szentes can produce good quality pellets from materials with 30 % of moisture content. During the production of agripellets, the incoming raw material first enters into the bale opener. The necessary grain size for pelleting is prepared by hammer mill. The manufacturing of pellets takes place in press machines at a high pressure and temperature. The cylindrical shape of pellets formed by the pressing machine mould can be flat or cyclic in shape. The diameter of the pellets are variable, the most common is the 6–8 mm size. Due to the high pressure and temperature, the lignin in the biomass is partially melted; it will hold together the particles after their release into the machine. The base material is inside the machine pressing mould, where the rotating mould seizes it and then in the inner mantle (0.5–1.00 mm), it is forced under closed rollers. Base materials are compressed by rollers in the peripheral surface of pressing machine mould into the appropriately trained bore, and hatching rods in the external mantle use a crushing knife to cut them to the correct size. It is of paramount importance in the production of high-quality pellets to coordinate the press hole and raw material. Figure 13.2 shows that the wall thickness of the pressing machine mould is broken into two different diameter bores. One of the bores

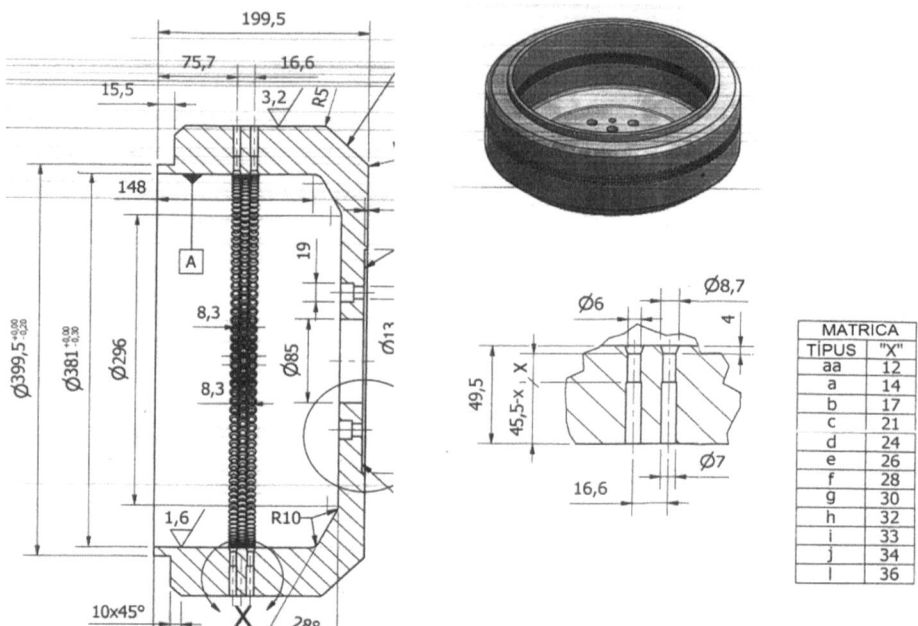

Fig. 13.2 Cross section of the pressing machine mould

in the inner mantle after the initial cone starts with 6 mm in diameter, which is called a compression hole, while the outer periphery has a 7-mm-diameter bore that goes against the press hole, and it is this bore that leads out. The bore length design depends on what kind of tree species (hard or soft) or straw pellets are being manufactured. The softer the species the longer the compression timber bore, while the shorter one for hardwood can be pressed into the right firmness [7].

13.4 Pelleting Rapeseed Stalk

In Hungary, in recent years, as a result of the manufacturing of biodiesel, the growing area of canola production has increased significantly. Cultivation is taking place on nearly 300 ha in the past few years. The straw generated, as by-product, is substantial, and varies between 3 and 6 t/ha. Most of the rapeseed stalk is returned to the soil after it is chopped into the ground but it is possible that some of the stubble is burned. Pellets were produced from rapeseed straw after grinding them with small pellet equipment and their energy characteristics were examined. At first, rapeseed stalks must be ground down to the correct size, which is done at a hammer crop mill. After grinding, good quality pellets were managed to be produced without any additional work or material. Pellets produced from rapeseed stalk are not fractured; these lengths were between 3.5–4 cm and 6 mm in diameter.

Table 13.1 Energetic characteristics of rapeseed stalk

	Moisture content W%	Heat value MJ/kg	Ash content AS%
Rapeseed stalk	12.5	16.0	5.1
Rapeseed pellets	11.5	16.2	5.1

Fig. 13.3 T&T Technik Ltd. agripellet plant, bale-opener, wheat straw before the grinding process

Straw and pellets were examined in an energetic laboratory. The calorific value, moisture content and ash content were determined; Table 13.1 shows the results.

The ash content is important energetically because the design of the combustion plant is essential. The ash content of wood pellets is low, less than 1%, while the ash content of herbaceous plant pellets is higher, about 3–10% (compare Table 13.1). The determination of moisture content is important because of the compressing; if it is too high or too low, the pellets will fall apart or crumble. The optimal moisture content is between 10 and 12%.

13.5   Presentation of the Studies

The location of examination was T&T Technik Ltd., an argipellet provider company, which operates in Szentes. They developed specific pellet production lines with which good quality different blend pellets from herbaceous plant materials are produced (see Fig. 13.3). A wide variety of agricultural by-products are utilized; among others, wheat straw, corn seed and rapeseed stalk, derivatives of oilseeds, waste from grain cleaning plants, as well as pellets that have been made from energy cane and energy grass.

Blend pellets are often produced in the factory. The laboratory tests were performed with two samples. The first sample contained a higher proportion (60%) of rapeseed stalk. In addition, the basic materials were corn stalk, wheat straw, and tailings. The second sample contained less rapeseed stalk (40%) and more wheat straw and tailings. Calorimetry measurements were performed on the two samples; also the ash contents were determined with a glow-furnace examination. Table 13.2 shows the results.

Table 13.2 Results of the examination of moisture content, heat values, and ash content

	Moisture content W%	Heat value MJ/kg	Ash content AS%
Sample 1	9.6	16.65	5.27
Sample 2	8.6	16.01	8.39

From the results of the experiment, we can conclude that sample 1 has a higher calorific value and lower ash content. The larger amount of rapeseed stalk causes higher heat value results in sample 1. The higher ash content of sample 2 may be caused by wheat straw and duckbill. The moisture content and heat value of the two examined samples can be considered a good value for agripellets. The ash content is also optimized for the different ash content of straw and crop residues, which are usually around 5–10%. Due to the higher ash content, it is required that special moving grate furnace is used when using agripellets. An important question is how much energy can we invest in the production process and how much energy is gained back. The use of energy varies depending on the basic material and the quantity produced per hour. The amount of pellets produce per hour is around 700–1000 kg. The energy consumption is 120–170 kWh. Most important specific energetic indicators are calculated by using the measured and calculated values. The balance of the energy efficiency is the energy content of product and primary energy input relative to 1 t of product. The energy efficiency is reduced with: energy intake of energy content of product/energy content of product × 100.

The energy demand in primary energy of the basic technology for 1 t product is 1713 MJ (476 kWh). Primary energy of 1 t of pellets from the basic production technology energy demand is 1713 MJ (476 kWh). From the 1st sample, the calculated recovered energy to the energy balance is 1:9.7, that is, ten times of the primary energy invested can be recovered. The energy balance is 1:9.4 if it is calculated using the heat value of sample 2. Expressed in degrees of energetic efficiency:

$$H = \frac{(Eo - Ei)}{Eo} \times 100 \, , \tag{13.1}$$

where H: energetic efficiency, Eo: energy output, Ei: energy input [8].

The value of the 1st sample is 89.7% and the 2nd sample is 89.3, which can be considered good values. The energy spent on delivery does not appear in this data. This would be a significant change in the energy balance because of the law of bulk raw materials. Therefore, efforts should be made for the local utilization of products.

13.6 Conclusions

Energy efficiency of wood pellet manufacturing was examined by our previous studies at Pellet Product Ltd., which operates in Petöháza [9]. The energy efficiency of production was 92.3%, which is a very good value thanks to the high heat value of pine shavings. In comparison, 89.5 % efficiency of agripellet production is a good value from an energetic point of view and well worth it. On the basis of laboratory tests, rapeseed stalk pellets have a high calorific value compared with herbaceous plants. Examining the whole process of cultivation and output energies, interesting conclusions were established. We examined how much energy per hectare can be obtained from the seeds of rapeseed and how much of the stalk. In the case of 2.5 t/ha, an average yield is approximately 60 GJ of energy from the seeds. When calculating with 3 t/ha, the utilizable quantities of rape stalk is 48 GJ, while calculating with 4.5 t/ha, it is 73 GJ energy [10]. It is noteworthy that often the same amount of energy left in arable lands such as rapeseed seeds can be obtained. In Hungary, there is a large potential for the utilization of agricultural by-products. The production of high-energy efficient agripellets can be an effective way to utilize the energy left in the fields.

References

1. BAI A. et al. (2002): *A biomassza felhasználása.* Szaktudás Kiadó Ház, Budapest.
2. TÁRMEG J. (2008): *Teendö a szármaradványokkal,* Agrárágazat, 9. évf. 9. Szám.
3. PANNONPELLET (2011) Jövö=Agripellet [online] Available from: http://www.pannonpellet. hu/publicistica.php?newsid=1127
4. CARBOROBOT (2012) *Pellet, biobrikett* [Online] Available from: http://www.carborobot.hu/ HU/Pellet.htm
5. BODRI, G. (2011) *Az agripellet megítélése Európában és Magyarországon.* Elöadás, Agro+-Mashexpo.
6. TÓVÁRI P. (2011) *Az agripellett energetikai és minöségi követelmény rendszere.* Elöadás, Agro+Mashexpo.
7. BURJÁN, Z. (2010) Pelletfütés II. Pelletgyártás-Víz- Gáz- Fütéstechnika áprilisi szám [Online] Avialable from: http://www.pannonpellet.hu/publicistica.php?newsid=978
8. SEMBERY, P., TÓTH, L. (2001) *Hagyományos és megújuló energiák. Szaktudás Kiadó Ház,* 260–261.
9. MAROSVÖLGYI, B., PAPP, V. (2010) *A pelletálás energiamérlegének vizsgálata.Tudományos eredmények a gyakorlatban,* Szolnok, 101–105.
10. PAPP, V., MAROSVÖLGYI, B. (2012) *A pellet mint megújuló energiahordozó elöállítása, hasznosítása és energetikai értékelése.* Energiagazdálkodás, 53. évf. 2:18–20.

Assessing the Availability of Biomass Residues for Energy Conversion: Promotors and Constraints

14

Johannes Lindorfer and Karin Fazeni

Abstract

The demand of biomass for energy production purposes will further increase over the coming years. For future attempts of implementing a sustainable energy system based on renewable energy carriers, agricultural residues are seen as a high-ranked option. This chapter presents background information for the biomass market in the European Union (EU), and the economic and ecological potential of straw as bioethanol feedstock is evaluated considering influencing factors for straw from being a residue to become a resource.

14.1 Introduction

The demand of biomass for energy production purposes will further increase over the coming years. According to the estimates, up to the year 2020, there will be a gap in biomass demand and supply which probably only can be met with imports. Forestry will remain the most important biomass feedstock followed by agricultural biomass for energy production purposes [1]. For future attempts of implementing a sustainable energy system based on renewable energy carriers, agricultural residues are seen as a high-ranked option as they show only small indirect land use change (ILUC) potential [2]. Especially the production of bioethanol from straw is highly promoted in Europe by various research and demonstration projects [3, 4]. There is still the need to examine the sustainability of bioethanol made from straw concerning greenhouse gas (GHG) emission savings and agro-environmental effects of alternative straw usage.

J. Lindorfer (✉) · K. Fazeni
The Energy Institute, Johannes Kepler University, Altenberger Street 69, Linz 4040, Austria
e-mail: lindorfer@energieinstitut-linz.at

© Springer Fachmedien Wiesbaden 2015
G. Dell, C. Egger (eds.), *World sustainable energy days next 2014,*
DOI 10.1007/978-3-658-04355-1_14

14.2   Biomass Resource Availability and Conversion Pathways

One of the biggest challenges for the biomass sector in the European Union (EU) is rea-ching the targets of the national renewable energy action plans (NREAP). For the year 2010, a domestic (European) availability of solid biomass and biogas of about 82 Mtoe was announced and a yearly growth rate of about 4.7% from 2015 to 2020 is estimated. The EU member states assume a primary biomass supply for their NREAPs at a value of 135 Mtoe. In contrast to that, a study conducted by Pöyry Energy Consulting assumes a biomass availability of about 120 Mtoe in the year 2020. As a result of different scenarios, there will be a supply gap of about 26–38 Mtoe which will increase the need for biomass imports. For the year 2020, it is projected that 71.4 Mtoe biomass from forestry, 36.3 Mtoe biomass from agriculture and about 13.9 Mtoe waste biomass are available in the EU for energy purposes. It is estimated that in the year 2020, 146–158 Mtoe solid biomass will be used in the heat and power sector within the EU [1].

14.2.1   Forestry Biomass

In the EU, around 27.6 Mm^3 swe pellets are produced and about 4.7 Mm^3 are imported. The exports in this sector are very small and account for 0.1 Mm^3. This results in a pellet consumption of about 32.1 Mm^3. Wood industry itself is a major consumer of wood in the heat and power plants [5].

Table 14.1 indicates that up to the year 2030, the demand for wood of the heat and power sector will make up about half of the total energetic wood demand [6]. In contrast to that, liquid biofuel production (Fischer–Tropsch Synthesis) from wood will stay cons-tantly low over time. Today, and also in the future, wood represents the most important biogenic energy source. For the expansion of the energy and material use of wood, it will be necessary to mobilize additional wood resources. To cover the excess demand,

Table 14.1 Status quo, projections of availability and demand of wood in the EU-27. (Source: Own table)

Year	Availability in Mm^3	Demand in Mm^3		Reference
		Total	For energetic use	
2005	754	779	332	Steirer [7]
	822	822	341	Mantau et al. [6]
2010	993	825	346	Mantau et al. [6]
	964	966	435	United Nations [8]
2020	1008	1145	573	Mantau et al. [6]
	1105	1064	504	United Nations [8]
2030	1109	1425	752	Mantau et al. [6]
	992–1421	1168–1419	585–859	United Nations [8]

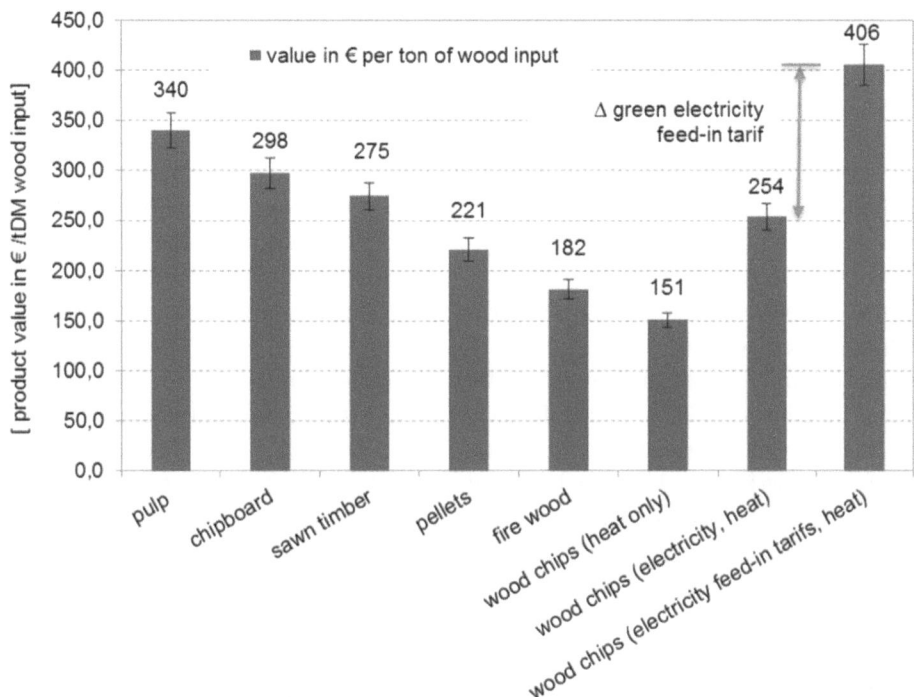

Fig. 14.1 Results for the direct supply chains of the material and energetic utilization of the raw material wood. (Source: Own calculation based on central European prices; Results normalized in euros per ton of wood input)

the expansion through the cultivation of fast-growing tree species, timber imports or the possible substitution by straw and other residues from agriculture and industry are under discussion. The prices for logs depend heavily on logging statistics, and thus the impact of damaging events. Production costs for wood chips vary according to supply-chain design from 5 €/bulk m³ to 14.6 €/bulk m³. It makes a big difference if the wood chips are produced from wood-harvesting residues or from trees harvested exclusively for chip production [9]. The price of wood products for energy usages is very country specific. The highest price for wood chips (20 or 30 % moisture content, particle size P 45 or P 31.5) was assessed in Austria at 143 €/t followed by Germany with 137 €/t and Ireland with 136 €/t [10]. In Fig. 14.1, the value chains of material and energetic utilization pathways of the raw material wood are compared.

The largest value is generated in the production of pulp for fibres in the investigated product field of semi-finished products (with consideration of the by-products like waste liquor) and green electricity with subsidized feed-in tariffs.

14.2.2 Agricultural Residues

Agriculture provides a wide range of renewable resources for different energy purposes (energy crops, residues such as manure and straw). There are usually several ways to render a particular biomass range. For dry lignocellulosic (woody) biomass, thermal or thermochemical (combustion, gasification) process options are utilized, while wet biomass with high protein or fat content is processed by biochemical conversion pathways (fermentation). As a consequence of the price developments and the food versus fuel dilemma, lignocellulosic materials like straw are investigated as raw materials for different conversion technologies.

Agricultural residues, such as cereal straw, are an interesting feedstock for future biofuel production as 56 % of the arable land in the EU is cultivated with cereals [11]. According to these uncertainties, the potential ethanol production from straw for the year 2020 ranges from 5.5 to 71 billion L in studies. Consequently, it is not possible to estimate accurate future production trends [12]. Latest developments in the EU biofuel policy suggest a ceiling for conventional biofuels produced from cereals and oil seeds and a forced promotion of advanced biofuels from agricultural residues, algae and bacteria [13].

14.3 From Residues to Resources: Case Study for Wheat Straw

14.3.1 Ecomomic Aspects of Conversion Pathway

Economical transportation and seasonal availability: The residues occur regionally distributed and consequently they must be extensively collected and transported. In particular, ash-rich biomass has a lower volumetric energy density (about 2 GJ m^{-3} for straw). Raw material storage is an element in the provision of raw materials for bridging the gap between logistics and subsequent production. For straw bales, approximately the 17-fold storage space is required compared to fuel oil based on the energy density [14]. In Fig. 14.2,

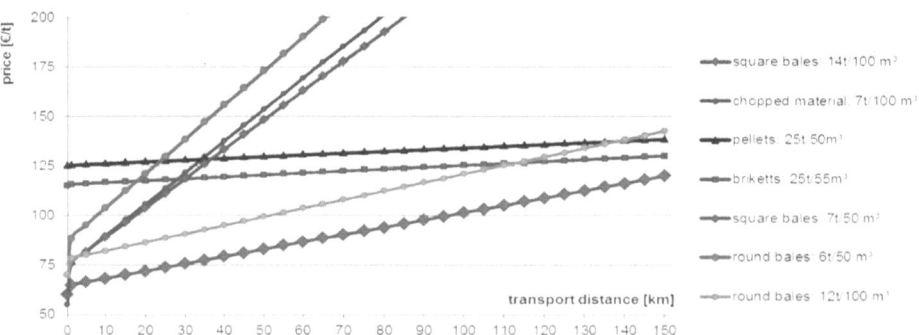

Fig. 14.2 Cost accounting for the raw material wheat straw as a function of distance and transport logistics. (Source: Own calculations based on [15, 16])

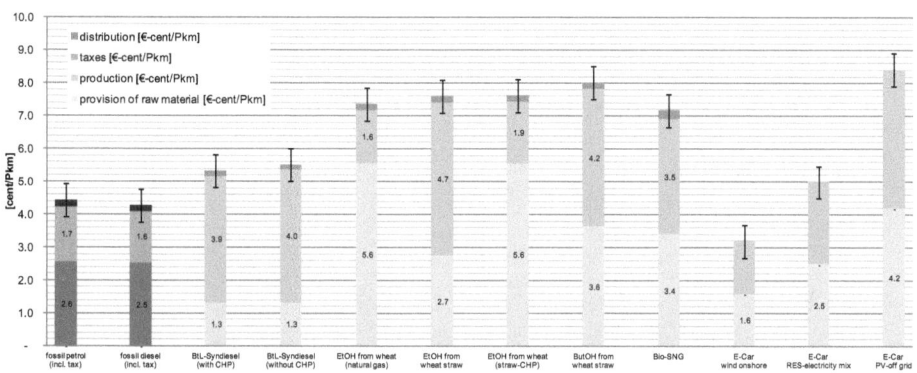

Fig. 14.3 Overview of the total specific cost per service unit (person kilometer, Pkm; excl. car). (Source: Own calculation and representation)

the dependencies of raw material logistics are illustrated in a cost estimate. The minimum purchase price for wheat straw can be reached by maximum loading of a truck with square bales.

Efficient conversion pathway: The selected conversion technology should be able to handle a wide range of feedstock with high conversion efficiency. However, the system size is often limited by the local heat loss on the one hand and the biomass logistics on the other hand. The basis of already-known technologies from the chemical processes can help to shorten development times.

The economic performance of various biofuel technologies from lignocellulosic raw materials is evaluated for comparative evaluation by calculation of specific costs of the service capacity, the kilometers traveled by an average vehicle and for the energy-equivalent biofuel proportion. The economic calculation started from a system life of 15 years and an interest rate of 5%. As a benchmark, the cost of fossil fuels and electric mobility is applied. The result in Fig. 14.3 is a benchmark evaluation comparable with other systematic reviews [17–19].

It has to be pointed out that resulting costs harmonize if costs for the passenger car, included as e-cars, are at the moment available at higer costs.

14.3.2 Ecological Aspects of the Straw to Bioethanol Conversion Pathway

As part of the EU, sustainability framework—the Renewable Energy Directive 2009/28/ EC (EU-RED)—includes a set of mandatory criteria and monitoring/reporting requirements concerning biofuel production and use [20]. In comparison to fossil fuels, a minimum of 35% GHG savings is obligatory, increasing to 50% in 2017 and 60% in 2018 for new biofuel plants. Second-generation biofuels would be double credited concerning the mandatory target [21]. Concerning biomass feedstock, the requirements for good agricul-

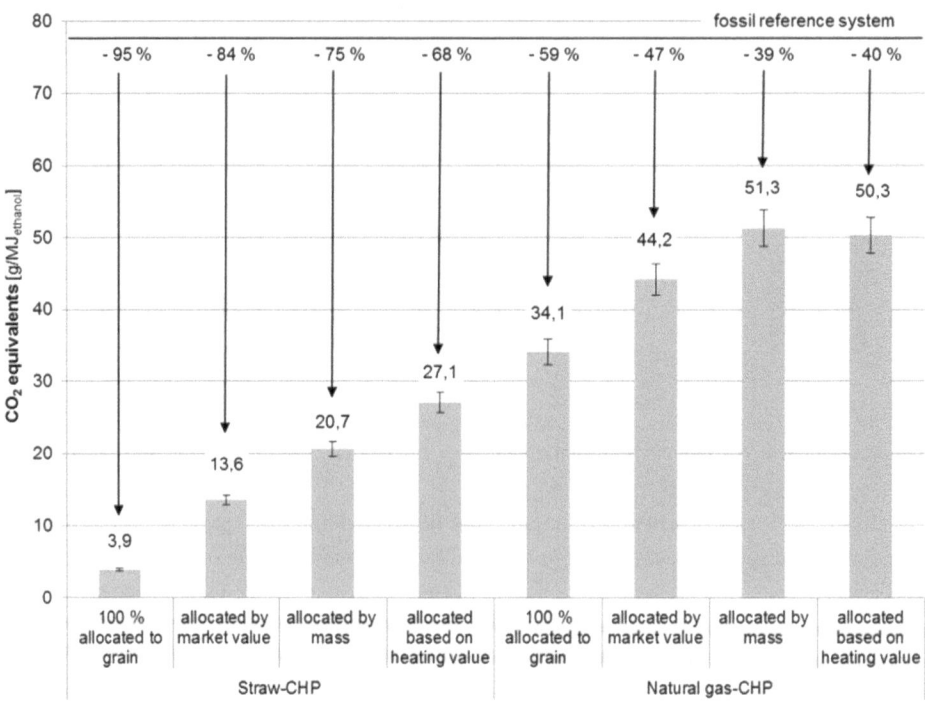

Fig. 14.4 Associated greenhouse gas emissions of the production of bioethanol from straw for different allocation scenarios. (Source: Own representation)

tural and environmental conditions are relevant and included within the EU across compliance rules of the common agricultural policy (CAP) [22]. Depending on the process energy source and allocation method, GHG emission savings from 39 to 95 % compared to fossil gasoline are achievable. These savings are realized if the agricultural inputs are allocated 100 % to the cereal grain and if a straw-fired cogeneration process provides the energy for the production of bioethanol. In this scenario, the GHG output expressed in CO_2 equivalents is 3.9 g/MJ$_{ethanol}$, which is by far the lowest value, compared to other bioethanol production routes (see Fig. 14.4). The allocation of all inputs on the grain is considered critical because of the possibility of overestimating the savings of bioethanol production from straw.

The majority of reported life-cycle assessment (LCA) or GHG balance studies on bioethanol from straw do not include ecological sustainability of raw material removal from arable land [23–25]. In LCA considerations, cereal straw is usually treated as a waste material that does not compete with other uses. However, cereal straw is already being used as a bedding material in animal husbandry and plays a major role for soil organic carbon reproduction. Thereby, soil organic carbon balance is an additional assessment tool for considering the compatibility of straw removal in regard to long-term maintenance of soil fertility [26, 27]. In the present study, the Verband Deutscher Landwirtschaftlicher Untersuchungs-und Forschungsanstalten (VDLUFA) method [28] is chosen for calculating the soil organic carbon balance as a basis for quantifying the recoverable amount of straw for energy conversion at the level of the German federal states. Particularly high levels of potentially usable energy straw, up to 50 %, are estimated for about a half of the examined federal states. When selecting a location for a bioethanol plant, the absolute available quantities of raw materials are a major decision parameter. Based on this evaluation in Germany, about 19 million t/a of straw are available for energy purposes. The possible recoverable amount of straw for energetic use is highly dependent on the prevailing agrarian structure in the relevant district. Additionally, it is important to check soil organic carbon balances on a smaller scale, ideally on the farm level, to gain accurate results. The retention of straw on the field ensures soil organic carbon reproduction, and is an important ecological constraint that must be considered in the calculation of raw material potentials for bioethanol production. An evaluation of the system, only applying LCA, is not sufficient for this goal because it does not take into consideration soil organic carbon balance and ecological raw material availability.

14.4   Conclusions: The Interrelation of Promotors and Constraints for Biomass Residues in Energy Conversion

Recently, the European Parliament decided to put a cap on the use of first-generation biofuels. As a result, a rapid switch to advanced biofuels made from agricultural residues or algae is favored. This is aimed to reduce the GHG emissions resulting from land-use changes. Currently, the parliament voted on such a draft legislation, which possibly will promote the market for advanced biofuels and lead to a stronger utilization of straw as bioethanol feedstock which can overcome major constraints of biomass availability for energy conversion [29].

Acknowledgments The support of this work by the association Energy Institute at the Johannes Kepler University and the funding under the Regional Competitiveness Programme 2007–2013 Upper Austria from the European Regional Development Fund and by the state of Upper Austria is gratefully acknowledged.

References

1. EURELECTRIC (2011) *Biomass 2020. Opportunities, Challenges and Solutions.* [Online] Available from: URL: http://www.eurelectric.org/media/26720/resap_biomass_2020_8-11-11_prefinal-2011-113-0004-01-e.pdf [Accessed: 07 October 2013].
2. SPOETTLE, M. et al (2013) Low ILUC potential of wastes and residues for biofuels. Straw, forestry residues, UCO, corn cobs. ECOFYS Project Number: BIEDE 13386/BIENL 12798.
3. BACOVSKY, D., DALLOS, M., WÖRGETTER, M., (2010) *Status of 2nd Generation Biofuels Demonstration Facilities in June 2010.* A Report to IEA Bioenergy Task 39 [Online] Available from http://www.task39.org/LinkClick.aspx?fileticket=PBlquceJcEQ%3D&tabid=4426&language=en-US [Accessed: 14 March 2013]

4. GNANSOUNOU, E. (2010) Production and use of lignocellulosic bioethanol in Europe: Current situation and perspectives. Bioresour. Technol. 101, 4842-4850.

5. MANTAU, U. (2012) *Wood Flows in Europe (EU 27).* Project report. Celle, 2012.

6. MANTAU, U. et al. (2010), EUwood—Real potential for changes in growth and use of EU forests. Final report, Hamburg, Deutschland.

7. STEIRER F. (2010) Current Wood Resources Availability and Demands. National and Regional Wood Resource Balances EU/EFTA Countries, United Nations, Geneva. http://www.unece.org/fileadmin/DAM/timber/publications/DP-51_for_web.pdf.

8. UNITED NATIONS (2011) The European Forest Sector Outlook Study II. http://live.unece.org/forests/outlook/welcome.html.

9. FRANCESATO, V. et al. (2008) *Wood Fuels Handbook. Production, Quality Requirements, Trading.* Italian Agriforestry Energy Association.

10. BIOMASS TRADE CENTER (2013) Monitoring of Wood Fuel Prices in Slovenia, Austria, Italy, Croatia, Romania, Germany, Spain and Ireland.

11. BENTSEN, N.S., Felby, C. (2012) Biomass for energy in the European Union—a review of bioenergy resource assessments. Biotechnology for Biofuels 5:25.

12. KRETSCHMER, B. et al. (2012) *Mobilising Cereal Straw in the EU to feed Advanced biofuel production.* Institute for European Environmental Policy.

13. EURACTIV (2013) *Biosprit: Neuausrichtung der EU-Agrarkraftstoffpolitik.* [Online] Available from: URL: http://www.euractiv.de/ressourcen-und-umwelt/artikel/biosprit-neuausrichtung-der-eu-agrarkraftstoffpolitik-007758 [Accessed: 09 October 2013]

14. RODE, M. et al. (2005) Naturschutzverträgliche Erzeugung und Nutzung von Biomasse zur Wärme- und Stromgewinnung', Bundesamt für Naturschutz, Bonn.

15. FNR (2005) ‚Leitfaden Bioenergie'.

16. KIESEWALTER, S. (2006) ‚Untersuchungen zur Verbrennung von halmgutartiger Biomasse'.

17. ZAH, R. et al. (2007) *Life Cycle Assessment of Energy Products: Environmental Assessment of Biofuels,* commissioned by the Swiss Federal Office for Energy, Environment and Agriculture.

18. EUROPEAN COMMISSION (2005) Viewls—*Clear views on clean fuels,* Project final report.

19. ARMSTRONG, A.P. et al. (2002) Energy and greenhouse gas balances of biofuels for europe, CONCAWE.

20. EUROPEAN COMMISSION (2009a) Directive 2009/28/EC of the European Parliament and of the Council of 23 April 2009 on the Promotion of the Use of Energy from Renewable Sources and Amending and Subsequently Repealing Directives 2001/77/EC and 2003/30/EC. European Commission, Belgium.

21. EUROPEAN COMMISSION (2009b) Directive 2009/30/EC of the European Parliament and of the Council of 23 April 2009 amending Directive 98/70/EC as regards the specification of petrol, diesel and gas-oil and introducing a mechanism to monitor and reduce greenhouse gas emissions and amending Council Directive 1999/32/EC as regards the specification of fuel used by inland waterway vessels and repealing Directive 93/12/EEC. European Commission, Belgium.

22. STOATE, C. et al. (2009) Ecological impacts of early 21st century agricultural change in Europe—A review. J. Environ. Manage. 91, 22–46.

23. BORRISON, A.L., McMANUS, M.C., HAMMOND, G.P. (2012) Environmental life cycle assessment of lignocellulosic conversion to ethanol: A review. Renew. Sustainable Energy Rev. 16, 4638–4650.

24. CHERUBINI, F. (2010) GHG balances of bioenergy systems—Overview of key steps in the production chain and methodological concerns. Renew. Energy 35, 1565-1573.

25. WILOSO, E.I., HEIJUNGS, R., DE SNOO, G.R. (2012) LCA of second generation bioethanol: A review and some issues to be resolved for good LCA practice. Renew. Sustainable Energy Rev. 16, 5295–5308.

26. SMITH, W.N., et al. (2012) Crop residue removal effects on soil carbon: Measured and inter-model comparisons. Agric. Ecosyst. Environ. 161, 27–38.
27. MUTH D.J. Jr., BRYDEN, K.M. (2013) An integrated model for assessment of sustainable agricultural residue removal limits for bioenergy systems. Environ. Model. Softw. 39, 50–69.
28. VDLUFA (Verband Deutscher Landwirtschaftlicher Untersuchungs-und Forschungsanstalten) (2004) *Humusbilanzierung: Methode zur Beurteilung und Messung der Humusversorgung von Ackerland.*
29. EUROPEAN PARLIAMENT (2013) *European Parliament backs switchover to advanced biofuels* [online] Available from: http://www.europarl.europa.eu/news/en/news-room/content/20130906IPR18831/html/European-Parliament-backs-switchover-to-advanced-biofuels [Accessed: 09 October 2013]

Subcritical Hydrothermal Liquefaction of Barley Straw in Fresh Water and Recycled Aqueous Phase

15

Zhe Zhu, Saqib Sohail Toor, Lasse Rosendahl and Guanyi Chen

Abstract

Barley straw was liquefied in fresh water and the aqueous phase obtained after lique-faction process at 300 °C. The effect of water recirculation on bio-oil yield and proper-ties was investigated. Results showed that bio-oil yield increased gradually to 38.4 wt % with the addition of recycled aqueous phase. The bio-oil contained similar functional groups and had higher heating values in the range of 27.29–29.34 MJ/kg. It showed that aqueous phase can be utilized as an effective solvent.

15.1 Background

Biomass is one of the most abundant and well-utilized sources of renewable energy in the world, which includes various natural and derived materials. The utilization of biomass can be environmentally friendly because it is a clean and carbon neutral resource [1]. Barley straw as an agricultural waste is abundant in Denmark. In 2012, 2.23 million t of straw was produced. However, only 24.67 % was used as energy source [2]. Therefore, it is necessary to make effective utilization of barley straw for energy production.

Z. Zhu (✉) · G. Chen
School of Environmental Science and Technology/State Key Laboratory of Engines,
Tianjin University, Weijin Road 92, Tianjin, China
e-mail: zhuzhezhz@163.com

S. S. Toor · L. Rosendahl
Department of Energy Technology, Aalborg University,
Pontoppidanstraede 101, Aalborg, Denmark

© Springer Fachmedien Wiesbaden 2015
G. Dell, C. Egger (eds.), *World sustainable energy days next 2014*,
DOI 10.1007/978-3-658-04355-1_15

121

Hydrothermal liquefaction (HTL) is a wet thermal conversion process, which is generally performed at lower temperature less than 400 °C in water or other organic solvent. It involves a series of thermal and chemical reactions, including hydrolysis, dehydration, decarboxylation, and repolymerization [3]. Water can act as both solvent and reactant in this process, so it can reduce the drying cost of raw material before use [4]. These biofuels are helpful in reducing the dependence on nonrenewable fossil fuel, improving our air quality and supporting rural economies.

Aqueous phase as one of the products was obtained after HTL, which contained a number of complicated compounds including acetic acid, alcohols, aldehydes, and pyran derivatives [5]. Stemann et al. [6] observed that the recirculated water phase obtained after hydrothermal carbonization of the poplar wood catalyzed the dehydration reactions. Considering the aqueous phase is rich in organic acids, it has the potential to be used as a reaction medium in HTL process. Based on the literature survey, there is little research on recirculation of aqueous phase in HTL process now. Thus, the subcritical HTL processes of barley straw with fresh water as well as aqueous phase obtained from the preceding experiment as solvent were studied in this chapter. The objective of this study is to evaluate the possibility of addition of aqueous phase in HTL of barley straw and investigate the effect of recirculation on products yield and properties.

15.2 Methodology

15.2.1 Materials

The raw material used in this study was barley straw, which was collected in Denmark. It was ground with a coffee grinder to particle size less than 1.0 mm before use. The analysis results of the barley straw were given in Table 15.1.

15.2.2 Hydrothermal Liquefaction and Separation of Products

The HTL experiments were carried out using a 1-L autoclave. In the reference experiment, 60 g barley straw, 400 g distilled water, and 10 wt% K_2CO_3 of the barley straw were loaded into the reactor. The reactor was sealed and purged with N_2 in order to remove air and prevent secondary reactions. Then, the reactor was heated to 300 °C and held for

Table 15.1 Analysis of barley straw

Proximate analysis			Ultimate analysis (wt%)						
Water content (wt%)	Ash content (wt%)	HHV (MJ/kg)	C	H	N	S	O	H/C	O/C
6.21	7.81	17.38	44.66	6.34	0.46	0.57	47.97	1.70	0.81

15 min. It was agitated at a speed of 300 rpm during the whole process. After the reaction is finished, the reactor was cooled to room temperature and then the gas was vented. The top phase inside the reactor was filtered through a pre-weighed filter paper under reduced pressure to obtain the aqueous phase. Then, the reactor and the stirrer were rinsed with acetone to recover the remaining oil/solid products. The mixture was centrifuged at 3000 rpm for 3 min to separate the acetone-soluble oil phase from insoluble phase. The supernatant acetone-soluble phase was further transferred into a pre-weighed evaporation flask and evaporated at 60 °C under a reduced pressure to remove acetone. The remaining dark liquid was weighed and referred to as bio-oil. The insoluble solid phase as well as the solid retained on the previous filter paper was oven dried at 105 °C for 24 h and weighted as solid residues.

In the recirculation experiments, it was carried out using the same method as in the reference run except that the addition of the aqueous phase.

15.2.3 Analysis of Products

Barley straw and bio-oil were analyzed by 2400 Series II CHNS/O element analyzer (PerkinElemer, USA) to determine the elemental contents. The composition of oxygen (O) was estimated by difference. Higher heating values of barley straw and bio-oil were measured using an IKA C2000 Basic Calorimeter. The analysis was performed twice, and the mean values were reported here. Thermogravimetric analysis (TGA) of all bio-oil was performed using PerkinElmer STA 6000. The sample was heated from 50 °C up to 950 °C at the heating rate of 10 °C/min. The TGA experiments were carried out under N_2 atmospheric with a flow of 20.0 ml/min. The FTIR spectra of bio-oil were recorded using an AVARTAR 370-FTIR spectrometer (Thermo Nicolet). The spectra range was set to be 4000–400 cm^{-1}.

Product yield was calculated based on the dry mass of initial feedstock used for HTL and defined as following equations:

$$\text{Bio-oil yield (wt \%)} = \frac{\text{weight of bio-oil}}{\text{weight of dried straw}} \times 100 \qquad (15.1)$$

$$\text{Solid residue yield (wt \%)} = \frac{\text{weight of solid residue}}{\text{weight of dried straw}} \times 100 \qquad (15.2)$$

$$\text{Conversion (wt \%)} = \frac{\text{weight of dried straw} - \text{weight of solid residue}}{\text{weight of dried staw}} \times 100. \qquad (15.3)$$

15.3   Results and Discussion

15.3.1   Effect of Recirculation on the Products Yield

The product distribution of bio-oil and solid residue obtained in the reference and recirculation runs is shown in Fig. 15.1. It can be clearly seen that the addition of recycled aqueous phase resulted in an increase of bio-oil yield from 34.9 to 38.4 wt% after three recirculation runs. GC–MS analysis of aqueous phase obtained from reference run indicated that it mainly consisted of various organic acids, polyols, and sugars, among which acetic acid and lactic acid were dominant species (data not shown). Therefore, the presence of organic acids may enhance the barley straw decomposition, increasing the bio-oil yield. Additionally, further conversion of other organic products existed in the aqueous phase may also contribute to the enhancement; however, it needs further verification.

Yield of solid residue significantly increased to 17.3 wt%, as compared to the yield of 8.1 wt% when the fresh distilled water was used. Accordingly, the biomass conversion decreased from 91.9 to 82.7 wt% with recirculation. Similar yield trend was observed in the hydrothermal carbonization of dry poplar wood chips by processing water recirculation [6], where the solid mass yield augmented. This can be explained by additional polymerization of reactive substances in the aqueous phase.

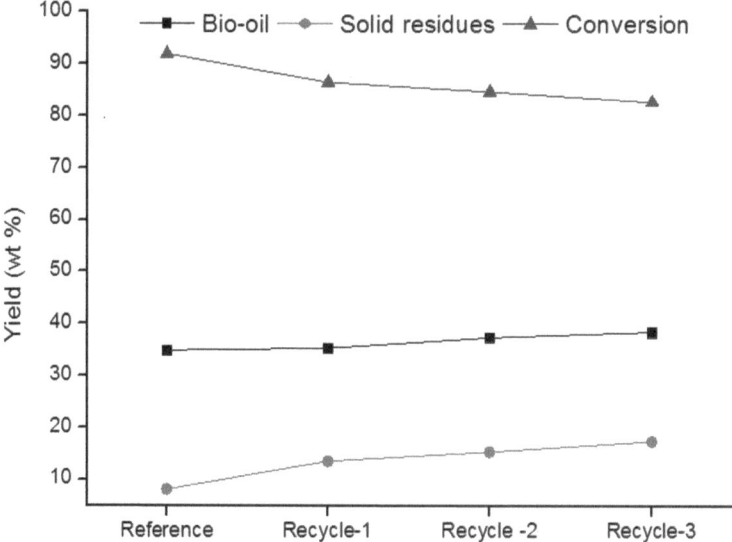

Fig. 15.1 Effect of recirculation on the product yield at 300 °C

Table 15.2 Elemental analysis of bio-oil from reference and recirculation experiment

Sample	Ultimate analysis (wt%)					H/C	O/C	HHV (MJ/kg)
	C	H	N	S	O			
Reference	67.89	7.62	0.75	0.56	23.18	1.35	0.26	27.29
Recycle-1	68.99	7.18	0.76	0.54	22.53	1.25	0.24	29.34
Recycle-2	68.55	6.94	0.7	0.52	23.29	1.21	0.25	28.66
Recycle-3	68.04	7.14	0.71	0.53	23.58	1.26	0.26	28.39

15.3.2 The Properties of Bio-Oil

15.3.2.1 Elemental Composition and Physical Properties of Bio-Oil

Elemental compositions of bio-oil from reference and recirculation experiments are given in Table 15.2. It can be seen that the carbon content was slightly augmented when the aqueous phase was recycled, which was probably due to the organic acids in the aqueous phase. It has been proved that the acid catalyst was found to be helpful for increasing the amount of bio-oil and carbon content [7]. The oxygen content of bio-oil was in the range of 22.53–23.58 wt%, which was much lower compared to raw biomass, indicating that oxygen was effectively removed in HTL. The H/C and O/C molar ratios of bio-oil were in the range of 1.21–1.35 and 0.24–0.26, respectively. The H/C molar ratio decreased in recirculation runs than that in the reference experiment, while the O/C molar ratio showed no great change. The higher heating values of bio-oil were in the range of 27.29–29.34 MJ/kg, which was significantly higher than that of barley straw (17.38 MJ/kg).

15.3.2.2 Thermal Stability of Bio-Oil

The TGA performed under inert atmosphere can be regarded as a miniature "distillation" and it can provide an estimated boiling point of bio-oil [8]. TG/DTG curves of bio-oil under N_2 are presented in Fig. 15.2. Heating the bio-oil to 950 °C resulted in mass loss, which was 76.13% in reference run and 71% in recirculation runs, indicating that the bio-oil obtained in HTL with the addition of recycled aqueous phase was less stable. The highest rate of weight loss of bio-oil was obtained in reference run and the first recirculation was observed at around 255 °C, while the second and third recycled runs had the maximum weight loss rate at around 305 °C.

Table 15.3 lists the boiling point distribution of the bio-oil obtained from different experiments. Obviously, all the bio-oil contained a large amount of high boiling point (>250 °C) materials. Only 15.09% of the bio-oil obtained in fresh water had a boiling point <250 °C and much of the bio-oil were of high molecular weight. In comparison, the addition of aqueous phase resulted in a slight increase in the fraction of lower boiling point materials, which was 16.80, 16.91, and 16.98%, respectively. It revealed that aqueous phase recirculation in HTL caused a slight increase in the fraction of bio-oil with a lower molecular weight.

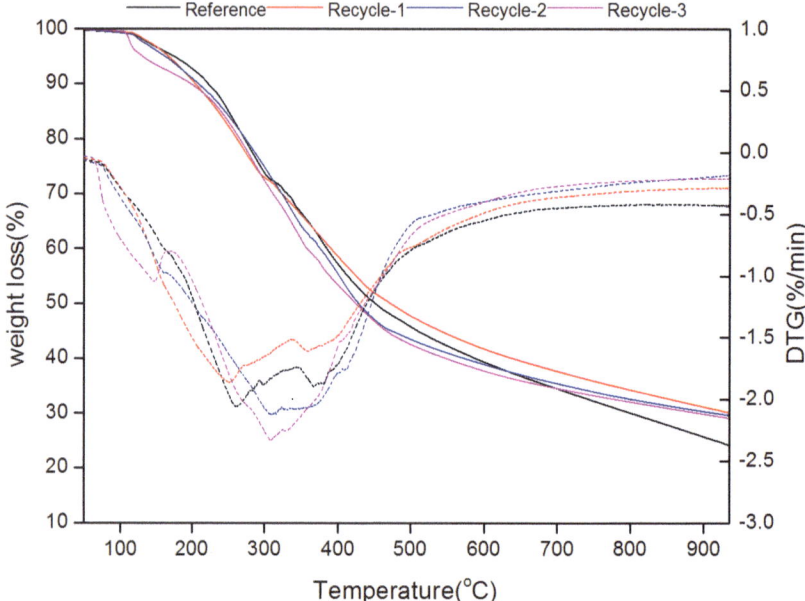

Fig. 15.2 Mass loss and derivative mass loss of bio-oil under N_2

15.3.3 FTIR Spectra of Bio-Oil

FTIR has been widely used to study the structure of bio-oil until now, since it can provide the information about certain components in the products through the absorption bands. The FTIR spectra of bio-oil obtained in reference and recirculation experiments are illustrated in Fig. 15.3. It can be observed that the spectra of bio-oil obtained from different solvents were similar, suggesting that they had the similar structure. The broad band in the 3700–3200 cm^{-1} region (peaked around 3330 cm^{-1}) was attributed to the stretching

Table 15.3 Boiling point distribution of bio-oil

Distillate range (°C)	Boling point of bio-oil			
	Reference	Recycle-1	Recycle-2	Recycle-3
50–200	7.28	9.36	9	10.17
200–250	7.81	7.44	7.91	6.81
250–300	11.13	9.29	10.23	10.7
300–350	7.31	6.9	10.8	10.7
350–400	9.4	7.64	8.82	8.33
400–450	6.9	6.53	8.18	6.52
450–500	4.41	4.18	3.6	4.23
500–550	3.52	3.39	2.47	2.68
>550	18.37	14.6	11.54	11.24

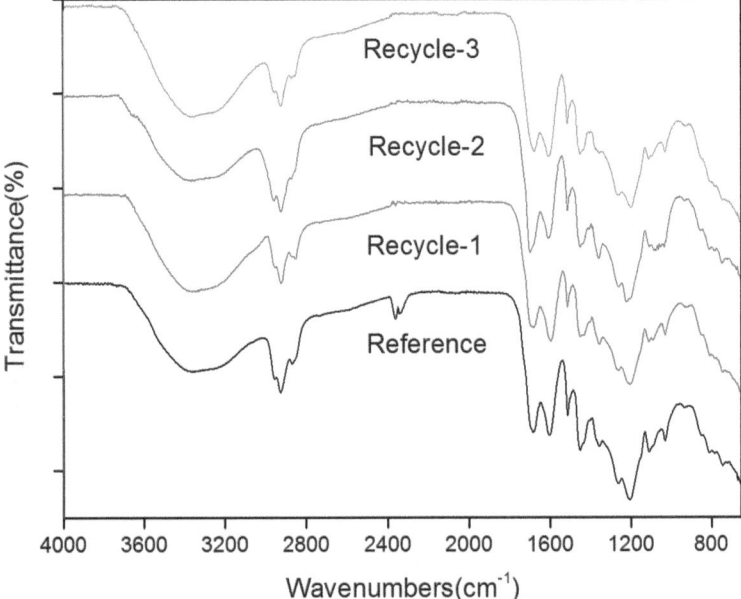

Fig. 15.3 FTIR spectra of bio-oil from reference and recirculation experiments

vibrations of alcoholic and phenolic OH groups [9], indicating that alcohols, phenolic compounds, and carboxylic acids were produced in the HTL process. The comparatively strong absorbance in the range of 2800–3000 cm^{-1} was caused by C–H stretch such as methyl and methylene groups [10], which provided the information that the aliphatic and alkyl aromatic compounds were present in bio-oil [11]. The absorption at 1686 cm^{-1} was ascribed to the presence of carbonyl groups in the form of ketones and aldehydes in the bio-oil. Also, the presence of both—OH stretching vibration and C=O stretching—revealed the presence of carboxylic acids. The absorption bands at 1604, 1514, and 1454 cm^{-1} related to the aromatic skeletal vibrations [12], together with the bands of aromatic C–H in-plane deformation at 1114, 1066, and 1034 cm^{-1} [9], suggested the existence of aromatic compounds in all bio-oils.

15.4 Conclusions

Barley straw was liquefied at 300 °C in fresh water and recycled aqueous phase, respectively. It was found that the bio-oil yield increased from 34.9 to 38.4 wt% with the increase of cycles. In the meantime, the yield of solid residues also increased. The addition of recycled aqueous phase was benefit for the enhancement of the carbon content and HHV of bio-oil. TGA results showed that the bio-oil obtained from recirculation runs contained lower boiling point materials compared with reference run. All the bio-oils have similar functional groups.

Further analysis of chemical composition of all aqueous phase and bio-oil will be conducted in the future in order to make a better understanding of the function of aqueous phase, which can guide the application in the industry scale.

Acknowledgment This work was financially supported by FLEXIfuel (DSF-BENMI Grant No. 10-094552). Zhe Zhu thanks the China Scholarship Council for the financial support.

References

1. RAGAUSKAS, A.J. et al. (2006) The path forward for biofuels and biomaterials. *Science*. 311(5760). 484–9.
2. DENMARK STATISTIC. (2013) *Straw yield and use by region, crop, unit and use*. Statistics Denmark. http://www.statistikbanken.dk/Statbank5a/SelectVarVal/saveselections.asp.
3. PETERSON, A.A. et al. (2008) Thermochemical biofuel production in hydrothermal media: A review of sub- and supercritical water technologies. *Energy & Environmental Science*. 1(1). 32–65.
4. KUMAR, S. (2013) Sub- and supercritical water technology for biofuels. In: LEE, J.W., editor. *Advanced biofuels and bioproducts*. New York: Springer.
5. SUN, P. et al. (2011) Analysis of liquid and solid products from liquefaction of paulownia in hot-compressed water. *Energy Conversion and Management*. 52(2). 924–33.
6. STEMANN, J., PUTSCHEW, A., ZIEGLER, F. (2013) Hydrothermal carbonization: Process water characterization and effects of water recirculation. *Bioresource Technology*. 143(0). 139–46.
7. YILGIN, M., PEHLIVAN, D. (2000) The properties of the oils obtained from liquefaction of poplar wood summary. *Turkish Journal of Engineering and Environmental Sciences*. 24(5). 305–13.
8. ROSS, A.B. et al. (2010) Hydrothermal processing of microalgae using alkali and organic acids. *Fuel*. 89(9). 2234–43.
9. COATES, J. (2000) Interpretation of infrared spectra, a practical approach. In: MEYERS, R.A., editor. *Encyclopedia of analytical chemistry*. Chichester: John Wiley & Sons, Ltd.
10. FAIX, O. (1992) Fourier transform infrared spectroscopy. In: LIN S.Y., DENCE C.W., editors. *Methods in lignin chemistry*. Berlin: Springer.
11. XIU, S. et al. (2010) Hydrothermal pyrolysis of swine manure to bio-oil: Effects of operating parameters on products yield and characterization of bio-oil. *Journal of Analytical and Applied Pyrolysis*. 88(1). 73–9.
12. FAIX, O. (1991) Classification of lignins from different botanical origins by FT-IR spectroscopy. *Holzforschung—International Journal of the Biology, Chemistry, Physics and Technology of Wood*. 45. 21.

Process Synthesis of Palm-Based Symbiotic Bioenergy Park

16

Rex T.L. Ng, Denny K.S. Ng and Raymond R. Tan

Abstract

This chapter presents a systematic approach for designing a palmbased symbiotic bioenergy park (SBP). In an SBP, material and energy exchanges among the processing facilities are facilitated to promote more sustainable operations in the palm oil industry. In this work, fuzzy optimisation is adapted to account for the individual economic interests of multiple parties in an SBP. The optimum network configuration which achieves the economic targets can be determined prior to detailed design.

16.1 Introduction

Palm oil accounted for 34.7% (58.77 million t) of global vegetable oil production (169.23 million t) in 2013/2014 [1]. With the increasing volume of palm oil production, large amounts of palm-based biomass (e.g., oil palm frond, empty fruit bunches, palm mesocarp fibre, palm kernel shell, etc.) are generated as by-products or wastes throughout the harvesting and milling processes. Palm-based biomass can be further converted into value-added products (e.g., biofuel, hybrid plywood, pellet, syngas, etc.) via various conversion technologies [2]. The concept of an integrated palm oil-based biorefinery (POB)

R. T.L. Ng (✉) · D. K.S. Ng
Department of Chemical and Environmental Engineering/Centre of Excellence for Green Technologies, The University of Nottingham, Malaysia Campus, Broga Road, Semenyih, Selangor 43500, Malaysia
e-mail: ngtonglip@gmail.com

R. R. Tan
Center for Engineering and Sustainable Development Research, De La Salle University, 2401 Taft Avenue, Manila 0922, Philippines

© Springer Fachmedien Wiesbaden 2015
G. Dell, C. Egger (eds.), *World sustainable energy days next 2014*,
DOI 10.1007/978-3-658-04355-1_16

has been introduced to integrate multiple biomass conversion technologies to produce various value-added products [3, 4]. Economic, environmental, and social aspects were considered simultaneously in synthesising a sustainable integrated biorefinery [5]. Recently, a new concept of palm oil processing complex (POPC) that consists of palm oil mill (POM), palm oil refinery (POR), POB and combined heat and power (CHP) was introduced to promote the interaction (e.g., mass and energy integration) of all processing facilities in palm oil industry [6].

Note that the above-mentioned works assume that the processing facilities are owned by the same company, and thus the individual economic performance of each facility is neglected. However, in reality, the processing facilities are normally owned by different owners. Thus, the concept of industrial symbiosis (IS) shall be included in promoting process integration within the processing plants that owned by different owners. According to Chertow [7], IS is the mutual exchange of materials and energy between industrial plants, specifically for enhancing sustainability. It can lead to synergies as energy, water, and materials exchange are involved in an IS system. Through the efficient utility sharing and material exchange among companies, greater economic benefit can be achieved. Based on IS concept, an eco-industrial park (EIP) is an industrial cluster which shares a common infrastructure that involves energy, water and materials. Kalundborg Park is one of the best examples of EIP. Its long history of operation demonstrates success in minimising material consumption and waste generation, while at the same time bringing financial benefits to participants by sharing resources [8]. However, the system emerged in an ad hoc manner, without the benefit of centralised, rigorous planning.

This paper presents a systematic approach for the synthesis and optimisation of a palm-based symbiotic bioenergy park (SBP), based on IS concept where interests of each processing plant are taken into consideration. Fuzzy optimisation is adapted to trade-off the predefined targets by each participating company independently. Optimal material and energy allocation network can be determined and companies with separate economic interests are partially satisfied within agreed upon bounds.

16.2 Problem Statement

The problem definition for the synthesis of a palm-based SBP may be stated as follows: Given a number of processing plants, $a \in A$ owned by different owners are interested in an IS scheme in sharing utilities and exchanging materials and energy among process plants. The major synthesis issue is to determine the optimum network such that the economic interests of the individual processing plants are satisfied simultaneously. In this chapter, the objective is to synthesise an integrated palm-based SBP that achieves maximum economic performance via fuzzy optimisation approach.

16.3 Problem Formulation

In this work, mass balance, energy balance and disjunctive constraint formulations are adapted based on the generic superstructure which is presented in [9]. For more detailed information about mass balance and energy balance formulations, it can be referred to the original work of Ng et al. [9].

Economic performance is evaluated based on the net present value of processing plant a (NPV^a) and it is expressed in the following equation:

$$NPV^a = \sum_{t}^{t_{max}} \frac{\left[GP^a \times \left(1 - \text{TAX}^a\right) + \text{DEP}^a \times \text{TAX}^a - \text{HEDGE}^a + \text{GOV}^a \right]}{\left(1 + \text{ROR}^a\right)^t} \quad \forall a \quad (16.1)$$

where GP^a is gross profit, TAX^a and DEP^a are the marginal tax rate and depreciation rate, respectively. HEDGE^a and GOV^a are expenses associated with hedging against catastrophic market actions and net benefits realised through governmental incentives or penalties, respectively. t_{max} is the operating lifespan and ROR^a is the expected rate of return.

In this work, fuzzy optimisation is adapted to satisfy the individual economic interest of each processing plant. An overall degree of satisfaction (λ) is introduced, which quantifies the degree of satisfaction of the least satisfied fuzzy goal [10, 11]. Each processing plant has predefined fuzzy goals for its performance, given by a linear membership function bounded by upper limit NPV^{aU} and lower limit NPV^{aL} as visualised in Fig. 16.1. As shown, λ approaches 1 as NPV^a approaches the upper limit and vice versa. The relationship between λ and targeted objectives in maximisation case is given as:

$$\lambda - \frac{NPV^a - NPV^{aL}}{NPV^{aU} - NPV^{aL}} \leq 0 \quad \forall a \quad (16.2)$$

Fig. 16.1 Fuzzy degree of satisfaction (λ) of the inequalities

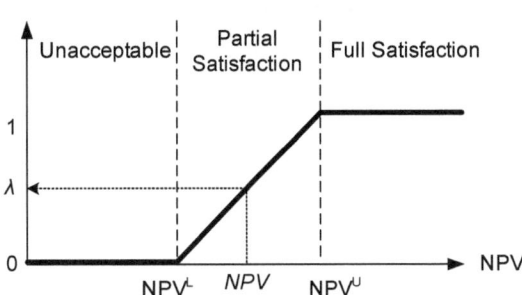

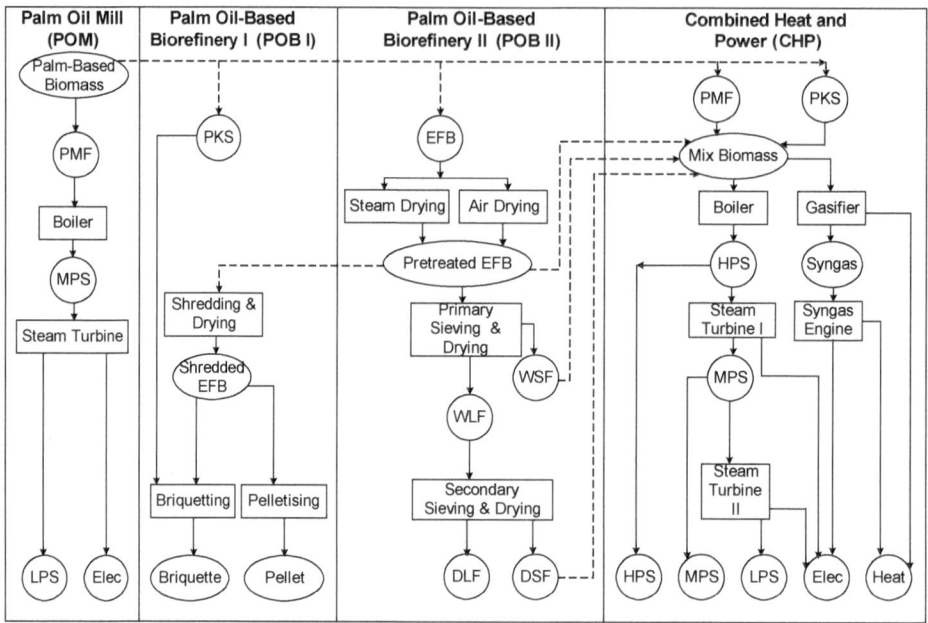

Fig. 16.2 Superstructure of case study [9]

16.4   Case Study

To illustrate the proposed approach, a palm-based SBP case study that is taken from Ng et al. [9] (see Fig. 16.2) is solved. As shown in Fig. 16.2, there are four processing plants interested in participating in the IS scheme. Palm-based biomass (e.g., empty fruit bunches (EFBs), palm mesocarp fibre (PMF), palm kernel shell (PKS), etc.) are identified as by-products generated from POM. Palm-based biomass can then be converted to pellet or briquette in POB I or dried long fibre (DLF) in palm oil-based biorefinery II (POB II). As shown, EFB needs to be pretreated with either steam or air drying systems prior to DLF production in POB II. In POB I, shredded EFB is produced from pretreated EFB and then further processed to produce pellets or briquettes. In briquette production, additional requirement of mixture of PKS and shredded EFB is needed.

The POM has an operating capacity of 60 t/h of fresh fruit bunches (FFBs) and all processing facilities are designed based on an operating lifespan of 15 years with annual operating time (AOT) of 8000 h. The price data of raw materials, palm green products, and energy, the mass conversion factor, economic data (including capital and operating costs) and energy consumptions (electricity and steam) for each technology from previous work [9] are adapted to develop this optimisation model.

The determination of upper and lower limits (NPV^{aU} and NPV^{aL}) will be first conducted and all upper and lower limits must be agreed by every participating processing plant.

Table 16.1 Economic data of case study

	Unit	POM	POB I	POB II	CHP
Upper limit, NPV^{aU}	million US$	12.50	7.86	12.95	6.49
Lower limit, NPV^{aL}	million US$	7.50	4.72	7.77	3.89
Net present value, NPV^a	million US$	12.50	5.33	8.79	6.48
Gross profit, GP^a	million US$/y	1.83	0.78	1.28	0.95
Payback period, PP^a	y	–	0.42	1.50	1.99

POM palm oil mill, POB I palm oil-based biorefinery I, POB II palm oil-based biorefinery II, CHP combined heat and power

It is assumed that NPV^{aU} of each processing plant are determined by getting maximum NPV values of each processing plant, while NPV^{aL} are determined by 40% lower than maximum NPV value of each processing plant. It is further assumed that each owner will be partially satisfied if their economic performance falls within the predefined upper and lower limits. The models with maximum NPV of each individual processing plant are first solved. The upper and lower limits results are summarised in Table 16.1.

Based on the predefined fuzzy limits, the optimisation objective is set to maximise the degree of satisfaction, λ, subject to Eq. 16.1 and Eq. 16.2 as well as mass balance, energy balance and disjunctive constraints adapted from Ng et al. [9]. The model is solved via LINGO v13.0 with Global solver in HP Compaq 6200 Elite Small Form Factor with Intel® Core™ i5-2400 Processor (3.10 GHz) and 4 GB DDR3 RAM. The global optimum solution is found and maximum λ is targeted as 0.195. Economic analysis of the case study solved is shown in Table 16.1. As shown, individual NPV of US$ 12.50, 5.33, 8.79, 6.48 million over 15 years lifespan for POM, POB I, POB II and CHP, respectively, are determined. Note that no additional capital investment of self-cogeneration plant in POM is considered because the POM imported steam and electricity from CHP, thus payback period of POM is zero.

The optimal network configuration is shown in Fig. 16.3. In this case, POM imports electricity and steam from CHP. All EFB are pretreated in steam drying and then sent to pellet production and DLF production in POB I and POB II, respectively. There is 2.23 t/h pretreated EFB sent to shredding, drying and pelletising processes to produce 1.26 t/h pellet. A total of 2.99 t/h DLF, 1.59 t/h WSF and 0.23 t/h DSF are produced throughout DLF production. Both WSF and DSF are sold to CHP and mixed with all PMF and PKS from POM before further combusted in boiler. Note also that 38.84 t/h HPS is produced from combustion in boiler and 3018 kW electricity are generated via both steam turbines I and II. CHP exported 810, 230 and 627 kW electricity to POM, POB I and POB II, respectively. The excess of 1351 kW of electricity can then be exported to the grid. After CHP exported steams based on requirement of POM, POB I and POB II, there is an excess of 0.38 t/h LPS can be exported to external facilities.

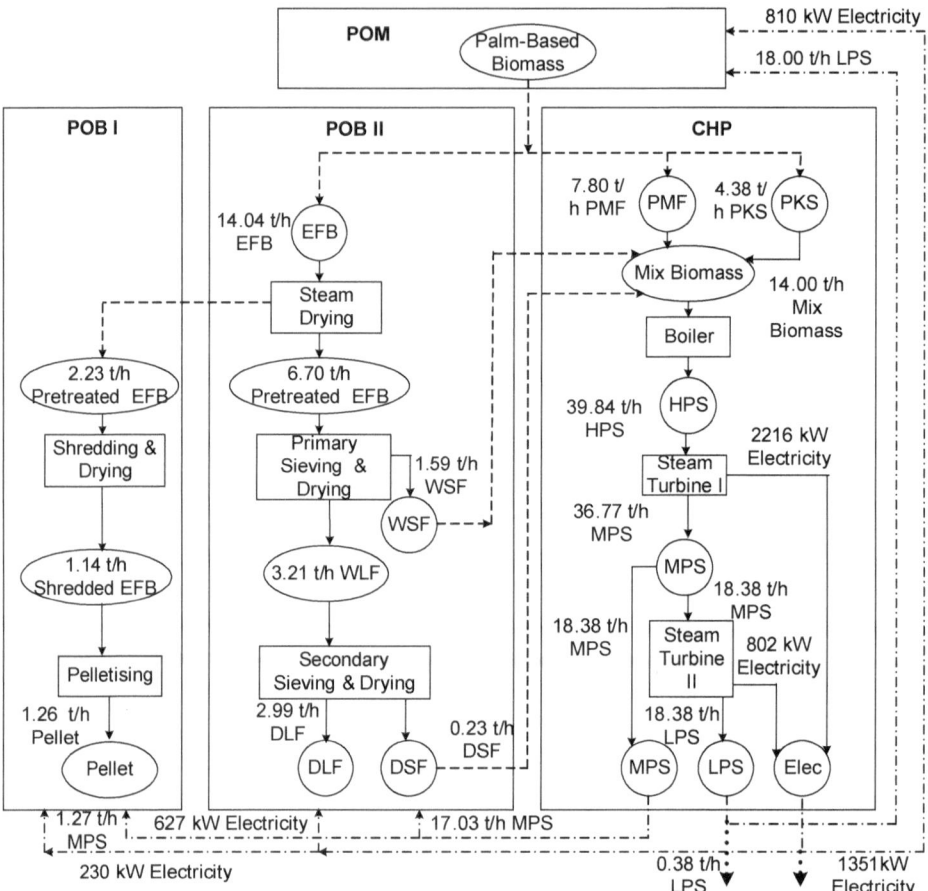

Fig. 16.3 Optimal configuration of case study

16.5 Conclusion

A systematic approach for synthesis of palm-based SBP is developed in this work. Fuzzy mathematical optimisation model is used to determine the detailed allocation of material and products that achieves the maximum economic performance of each processing plant simultaneously; the benefit of each owner is considered to determine a mutually acceptable symbiosis scheme. This approach thus makes the likelihood of successful implementation of IS more likely by taking into account the equitable satisfaction of the self-declared profit goals of the different parties. The proposed approach can be revised easily and adapted into different SBPs other than palm oil industry that involve multiple owners. Detailed assessment with the consideration of environmental, social and economic aspects can be included in the future works to synthesise a sustainable SBP.

Acknowledgment The financial support from Global Green Synergy Sdn. Bhd., Malaysia and University of Nottingham Research Committee through New Researcher Fund (NRF 5021/A2RL32) are gratefully acknowledged. In addition, authors would also like to acknowledge financial support from Ministry of Higher Education, Malaysia through LRGS Grant (project code: 5526100) and Commission on Higher Education PHERNet Sustainability Studies Program, Philippines.

References

1. UNITED STATES DEPARTMENT OF AGRICULTURE (2014) *Oilseeds: World Markets and Trade.* [Online] Available from: http://apps.fas.usda.gov/psdonline/circulars/oilseeds.pdf [Accessed: 8 June 2014].

2. NG, D.K.S. and NG, R.T.L. (2013) Applications of process system engineering in palm-based biomass processing industry. *Current Opinion on Chemical Engineering.* 2(4). 448–454.

3. KASIVISVANATHAN, H. et al. (2012) Fuzzy optimisation for retrofitting a palm oil mill into a sustainable palm oil-based integrated biorefinery. *Chemical Engineering Journal.* 200–202. 697–709.

4. NG, R.T.L., TAY, D.H.S. and NG, D.K.S. (2012) Simultaneous process synthesis, heat and power integration in a sustainable integrated biorefinery. *Energy Fuels.* 26(12). 7316–7330.

5. NG, R.T.L., HASSIM, M.H., NG, D.K.S. (2013) Process synthesis and optimisation of a sustainable integrated biorefinery via fuzzy optimisation, *AIChE Journal.* 59(11). 4212–4227.

6. NG, R.T.L. and NG, D.K.S. (2013). Systematic approach for synthesis of integrated palm oil processing complex. Part 1: Single owner, *Industrial & Engineering Chemistry Research.* 52(30). 10206–10220.

7. CHERTOW, M.R. (2004) Industrial symbiosis. *Encyclopedia of Energy.* 3. 407–415.

8. JACOBSEN, N.B. (2006) Industrial Symbiosis in Kalundborg, Denmark: A quantitative assessment of economic and environmental aspects. *Journal of Industrial Ecology.* 10(1–2). 239–255.

9. NG, R.T.L. et al. (2014) Disjunctive fuzzy optimisation for planning and synthesis of bioenergy-based industrial symbiosis system. *Journal of Environmental and Chemical Engineering.* 2(1). 652–664.

10. BELLMAN, R. and ZADEH, L. A. (1970) Decision making in a fuzzy environment. *Management Science.* 17B. 141–164.

11. ZIMMERMANN, H.J. (1978) Fuzzy programming and linear programming with several objective functions. *Fuzzy Sets and Systems.* 1. 45–55.

Biofuel from Lignocellulosic Biomass Liquefaction in Waste Glycerol and Its Catalytic Upgrade

17

Miha Grilc, Blaž Likozar and Janez Levec

Abstract

Liquid biofuel was obtained by solvolytic liquefaction of lignocellulosic (LC) biomass in glycerol. Different types of wood, reaction temperatures and homogeneous catalysts were tested to obtain solvolytic oil with suitable fuel properties. Liquefaction conditions have little effect on fuel properties, so the upgrade step had to be introduced. Feasibility of hydrodeoxygenation (HDO) upgrading method using commercially available Ni, NiMo and Pd catalysts was investigated under elevated pressure and temperatures between 200 and 325 °C.

17.1 Introduction

Lignocellulosic (LC) biomass is a ubiquitous renewable energy source with often unexploited potential due to limited applicability especially for transportation purposes. Forestry residues, sawdust from wood-processing industries or waste wood are often utilised only by burning on-site or even treated as waste and disposed. Glycerol, on the other hand, is a by-product of biodiesel production of which world production is above million tons

B. Likozar (✉) · M. Grilc
Laboratory of Catalysis and Chemical Reaction Engineering, National Institute of Chemistry, Hajdrihova 19, Ljubljana 1000, Slovenia
e-mail: miha.grilc@ki.si

B. Likozar · J. Levec
Laboratory of Catalysis and Chemical Reaction Engineering,
Faculty of Chemistry and Chemical Technology, University of Ljubljana,
Aškerčeva 5, Ljubljana 1000, Slovenia

© Springer Fachmedien Wiesbaden 2015
G. Dell, C. Egger (eds.), *World sustainable energy days next 2014,*
DOI 10.1007/978-3-658-04355-1_17

per year and by the year 2020 the estimated production will be six times over its demand [1]. The solvolysis of LC biomass is known to be effective in acidified glycols even at ambient pressure and temperatures below 180 °C, which is a promising alternative to high energy-consuming thermal processes like pyrolysis (over 450 °C) or gasification (over 700 °C). The liquefaction of wood is caused by depolymerisation and solubilisation of macromolecules of cellulose, hemicellulose and lignin. High conversion of wood into a liquid product within a reasonable time and low energy input is reported in the literature [2]. Liquefied biomass was successfully tested directly as a fuel in gas turbines and heating boilers, while unsuitable for use in internal combustion engine (e.g. diesel engine) due to high viscosity [3]. Calorific value is relatively high for biomass-derived oils although it is still much lower in comparison to conventional transportation fuels (diesel, gasoline) due to high content of chemically bonded oxygen.

Wood liquefaction in glycerol provides a cheap and simple possibility to produce second-generation biofuel from cheap renewable feedstocks that are generated in excessive quantities with respect to the market demand. Less than $0.5\,\mathrm{kW\,kg^{-1}}$ energy input required [2] and nearly zero price of the feedstocks on the world market could make this environmentally friendly process also economically attractive. Fuel properties can be improved by a catalytic hydrodeoxygenation (HDO) process, which is costly but technically feasible. Some advances of both processes are presented in this work.

17.2   Methods and Procedures

Conversion of LC biomass into quality liquid fuel consists of two consecutive steps:

- Liquefaction of solid biomass by solvolysis, using acidified polyols
- Catalytic upgrade (HDO) of liquefied biomass

Liquefaction by solvolysis is a reaction between a solvent and the macromolecules of LC biomass, which results in depolymerisation of cellulose, hemicellulose and lignin, and subsequent dissolution of soluble depolymerised products in residual solvent mixture. Typical reaction of cellulose solvolysis by acidified glycol is depicted in Fig. 17.1 and results in the formation of corresponding glucosides and further to levulinates [4]. The hemicellulose liquefaction mechanism is similar, while depolymerisation of lignin is more complex and might be responsible for repolymerisation reactions if liquefaction time is too long. Reactions are catalysed by mineral acids (sulphuric acid), organic acids (p-toluenesulfonic acid) and some Lewis acids (ionic liquids).

Subsequent catalytic HDO step is introduced to increase fuel calorific value by means of reduction of chemically bonded oxygen in liquefied biomass. HDO takes place at elevated temperature and pressure in the presence of heterogeneous catalyst and hydrogenating agents: hydrogen and a hydrogen donor solvent. HDO corresponds to hydrode-

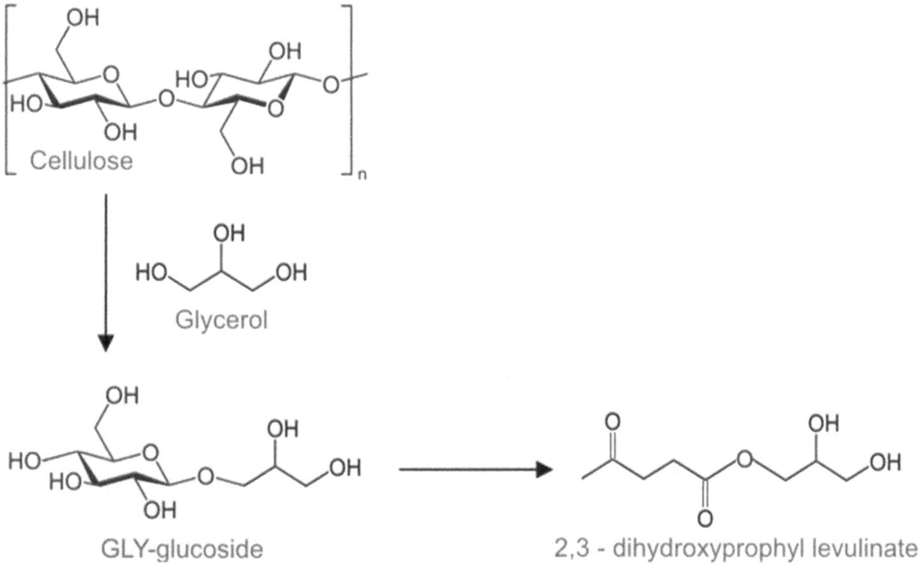

OH

Cellulose

OH

OH
HO
OH

Glycerol

OH

HO
HO
OH

GLY-glucoside

2,3 - dihydroxyprophyl levulinate

Fig. 17.1 Reaction pathway of cellulose solvolysis by acidified glycerol

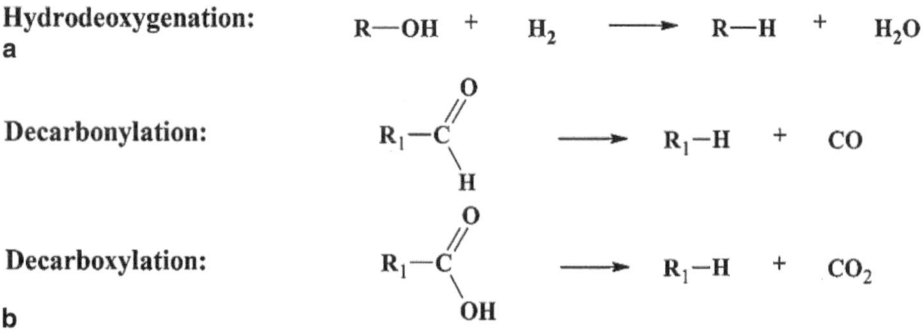

Hydrodeoxygenation:
a

$$R-OH \; + \; H_2 \longrightarrow R-H \; + \; H_2O$$

Decarbonylation:

$$R_1-\overset{O}{\underset{H}{\overset{\|}{C}}} \longrightarrow R_1-H \; + \; CO$$

Decarboxylation:

b

$$R_1-\overset{O}{\underset{OH}{\overset{\|}{C}}} \longrightarrow R_1-H \; + \; CO_2$$

Fig. 17.2 Main reaction of oxygen removal at catalytic upgrade of biomass-derived oil

sulphurisation process in crude oil refineries, that means, same commercially available catalysts (e.g. $NiMo/Al_2O_3$ or $CoMo/Al_2O_3$) can be used in the HDO process. Figure 17.2 shows most important reactions for oxygen removal from liquefied biomass; oxygen can be removed in the form of water, carbon monoxide and carbon dioxide. HDO is a catalytic process where hydrogen and heterogeneous catalyst are required. Decarbonylation and decarboxylation are homogeneous reactions caused by thermal decomposition, so the presence of hydrogen or catalyst is not strictly necessary. HDO is the most-desired deoxygenation reaction, as no carbon atom is removed, which results in a higher yield of the final product.

17.3   Experimental

All experiments were performed in a 300-mL stainless steel autoclave, equipped with
the magnetically driven Rushton turbine impeller of 30 mm diameter. Batch liquefaction
experiments were studied at temperatures between 150 and 200 °C and normal pressure
of nitrogen or air; LC biomasses of dried beech, spruce, oak or fir sawdust were tested.
HDO experiments were carried out at temperatures of 200–325 °C and hydrogen pressure of 2–8 MPa was maintained constant throughout each experiment. Gaseous phase
was continuously analysed, while liquid phase samples were collected periodically and
analysed later.

For the LC biomass liquefaction experiments, dry biomass and glycerol were placed
into reaction vessel in 1:3 mass ratio, p-TSA catalyst was then added (0–3 wt%). Reactor
was closed, agitation and heating program was started and the experiment was conducted
for 60–90 min at selected final temperature.

For the HDO reaction studies, solvolytic oil was provided by an industrial company
that runs a pilot reactor for wood liquefaction. Preparation procedure and product analysis
can be found in the literature [3]. Solvolysed oil was diluted with tetralin (3:1 mass ratio),
as it is known to have positive effect on the transformation of molecular hydrogen into
more reactive radicals. Commercially available Ni, NiMo and Pd catalysts were selected
for hydrotreatment. Catalyst loaded in the reaction mixture was 7.5 % per mass of the liquefied wood. Experimental setup for HDO operation mode is depicted in Fig. 17.3.

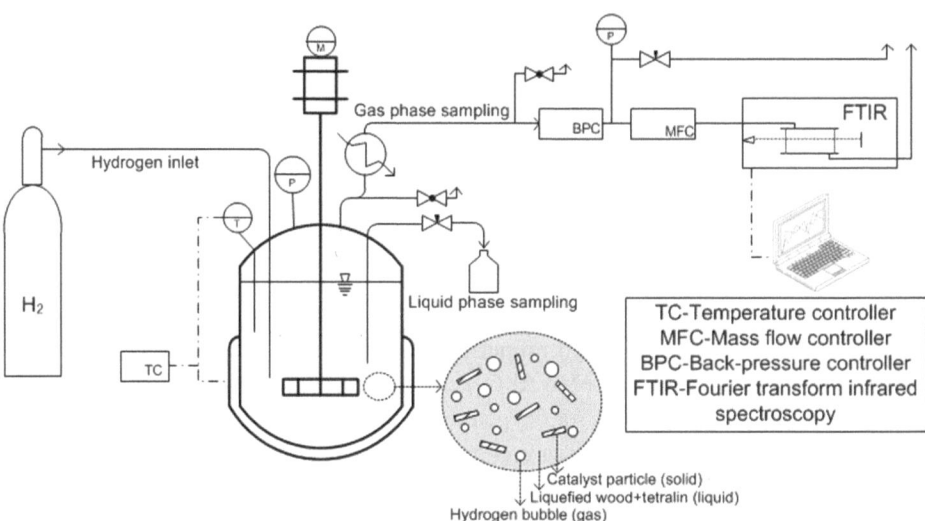

Fig. 17.3 Scheme of the reactor for catalytic hydrodeoxygenation of liquefied wood

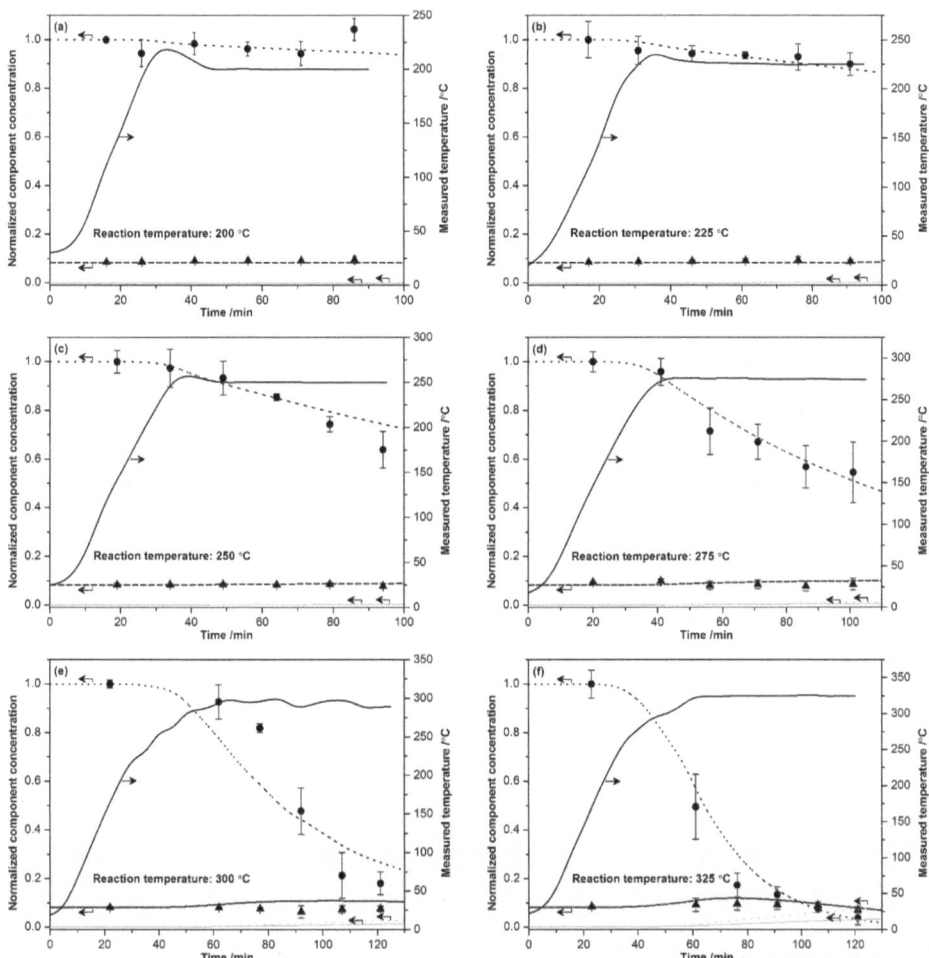

Fig. 17.4 Experimental (● OH group; ▲ C=O group) and modelling (·····, OH group; – – –, C=O group; ——, CO; ·····, CO_2) hydrodeoxygenation (HDO) results over $NiMo/Al_2O_3$ at temperatures: 200 °C (**a**), 225 °C (**b**), 250 °C (**c**), 275 °C (**d**), 300 °C (**e**), 325 °C (**f**) [5]

17.4   Results

During the liquefaction of beech wood by acidified glycerol, the formation of carbonyl functional group was detected by Fourier transform infrared spectroscopy (FTIR) analysis of the liquid phase. The peak detected at 1740 cm^{-1} indicates the liquefaction of cellulose in glycerol and resulting formation of esters (levulinates), according to mechanism presented in Fig. 17.1. When temperature plateau at 175 °C was reached, liquefaction reaction started, while between 60 and 90 min the levulianates formation rate was noted to decrease, as nearly all wood particles had already been solvolysed.

Liquefied wood had the oxygen content around 41 wt%, most of it in the form of hydroxyl functional group (28 mmol g^{-1}) and minor part (2 mmol g^{-1}) of carbonyl functional group. Figure 17.4 shows the influence of the reaction temperature between 200 and

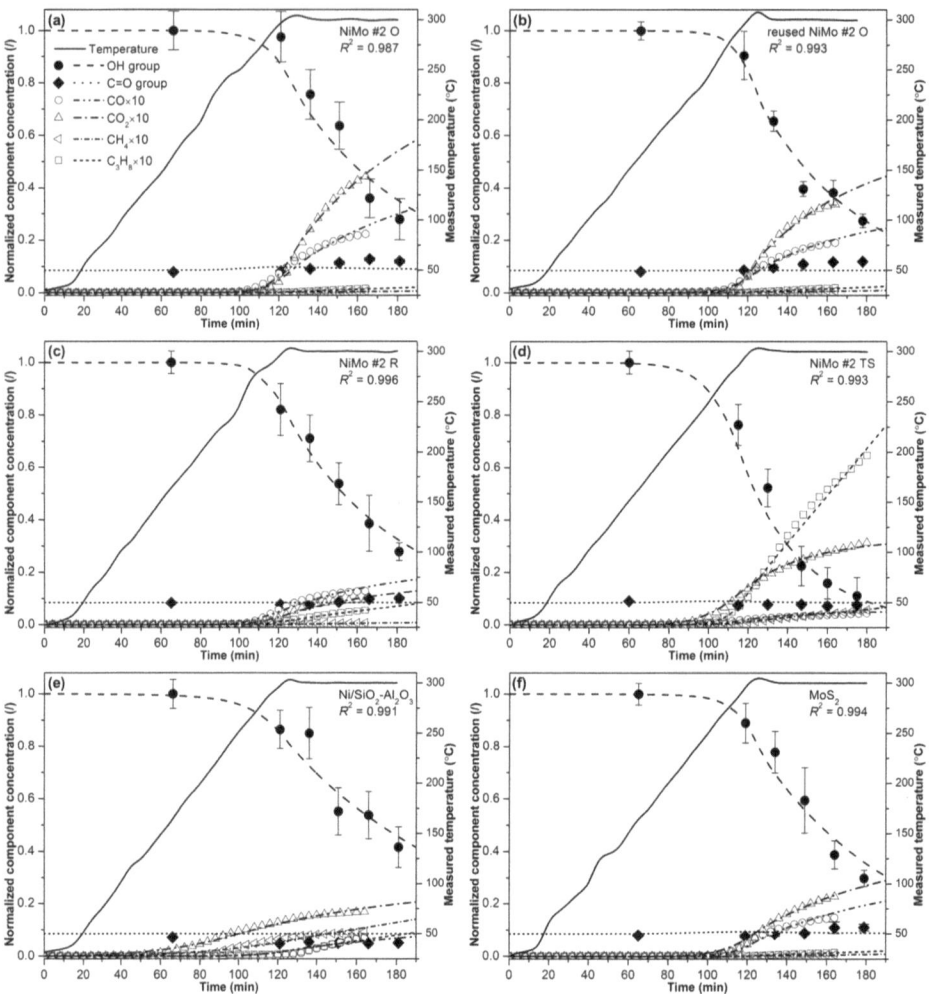

Fig. 17.5 Influence of NiMo in oxide (**a**, **b**), reduced (**c**), sulphide (**d**) form, Ni/SiO$_2$-Al$_2$O$_3$ (**e**) and bulk MoS$_2$ (**f**) catalysts on hydrodeoxygenation (HDO) of solvolysed biomass [6]

325 °C on the progress of oxygen (OH and C=O groups) removal from the liquefied wood at 6 MPa pressure of H$_2$, stirring speed of 750 min^{-1} and heat-up rate of 7.5 K min^{-1} [5].

Temperature of at least 300 °C is required to obtain a product with oxygen content below 8 wt% within 60 min, while NiMo catalyst in sulphided form was found to be most active and selective towards HDO in catalyst-screening study at 300 °C, as shown in Fig. 17.5 [6].

Chemically bonded oxygen was mostly removed from hydroxyl group, while carbonyl group was very resistant to HDO at present reaction conditions. Oxygen from hydroxyl group was removed in the form of water, which confirms high HDO selectivity, while decarbonylation and decarboxylation were slow even at high temperatures and with most active catalysts. Pretreatment of NiMo catalysts had significant effect on the selectivity of

Fig. 17.6 Dry sawdust (18 MJ kg^{-1}), liquefied wood before (22 MJ kg^{-1}) and after catalytic upgrade (up to 36 MJ kg^{-1})

all deoxygenation reactions. Figure 17.5 indicates that viscosity of upgraded solvolysed biomass is significantly lower, since NiMo/Al$_2$O$_3$ possesses some hydrocracking activity [6]. HDO method was found effective in improving fuel properties of liquefied biomass as gross calorific value of the liquefied wood increased from 22 to 36 MJ kg^{-1}, while viscosity significantly decreased (Fig. 17.6).

However, HDO is an expensive method, as hydrodeoxygenated pyrolysis oil costs more than US$ 700/t of oil equivalent [7]. Price for hydrodeoxygenated liquefied wood is expected to be much lower, as pyrolysis is significantly less effective (lower mass yield of the oil phase) and more energy consuming (higher temperature of the process) in comparison to LC biomass liquefaction in glycerol.

17.5 Conclusions

- Wood liquefaction in glycerol is efficient, cheap and environmentally friendly
- Extent of solvolysis can be followed by carbonyl group formation with FTIR
- HDO technically feasible but expensive process
- Temperature of at least 300 °C is required for high conversion in reasonable time
- Sulphided NiMo/Al$_2$O$_3$ possessed highest HDO and hydrocracking activity among tested catalysts.

References

1. CHRISTOPH, R., SCHMIDT, B., STEINBERNER, U., DILLA, W., KARINEN, R. (2000) Glycerol. In: *Ullmann's Encyclopedia of Industrial Chemistry*. Wiley-VCH Verlag GmbH & Co. ISBN: 9783527306732. doi:10.1002/14356007.a12_477.pub2. *URL: http://dx.doi. org/10.1002/14356007.a12_477.pub2*
2. KUNAVER, M., JASIUKAITYTĖ, E., ČUK, N. (2012) Ultrasonically assisted liquefaction of lignocellulosic materials. *Bioresource Technology*. 103 (1). 360–366.
3. SELJAK, T. et al. (2012) Wood, liquefied in polyhydroxy alcohols as a fuel for gas turbines. *Applied Energy*. 99 (0). 40–49.
4. YAMADA, T., ONO, H. (2001) Characterization of the products resulting from ethylene glycol liquefaction of cellulose. *Journal of Wood Science*. 47 (6). 458–464.
5. GRILC, M., LIKOZAR,B., LEVEC, J (2014) Hydrotreatment of solvolytically liquefied lignocellulosic biomass over NiMo/Al2O3 catalyst: Reaction mechanism, hydrodeoxygenation kinetics and mass transfer model based on FTIR. *Biomass and Bioenergy*. 63 (0). 300–312.
6. GRILC, M., LIKOZAR, B., LEVEC, J(2014) Hydrodeoxygenation and hydrocracking of solvolysed lignocellulosic biomass by oxide, reduced and sulphide form of NiMo, Ni, Mo and Pd catalysts. *Applied Catalysis B: Environmental*. 150–151 (0). 275–287.
7. MORTENSEN, P.M. et al. (2011) A review of catalytic upgrading of bio-oil to engine fuels. *Applied Catalysis A: General*. 407 (1–2). 1–19.

H$_2$ from SERP: CO$_2$ Sorption by Double-Layered Hydroxide at Low and High Temperatures

18

F. Micheli, L. Parabello, L. Rossi, P. U. Foscolo, K. Gallucci

Abstract

Sorption-enhanced processes of removing CO2 maximize hydrogen production. In this work, three kinds of CO2 sorbents were synthesized using a low supersaturation method and keeping the $M^{2+}/M^{3+} = 2/1$ (M^{2+} = Mg, Ca and M^{3+} = Al). High- and low-temperature capture tests were carried out in a fluidized bed reactor and compared with thermogravimetric analysis. Differential thermal analysis was used to investigate the sorption behavior. Samples were characterized by means of Brunauer–Emmett–Teller (BET)Barrett–Joyner–Halenda (BJH), X-ray diffraction (XRD), scanning electron microscopy (SEM)energy-dispersive X-ray spectroscopy (EDX) analysis.

18.1 Introduction

Renewable sources are the base to produce environmentally friendly H$_2$ that is currently produced mainly by catalytic methane steam reforming. Among different renewable production methods, biomass gasification can be considered an excellent process for the future "zero-emission" hydrogen production [1].

F. Micheli (✉) · L. Parabello · P. U. Foscolo · K. Gallucci
Department of Industrial Engineering, University of L'Aquila, 67100 L'Aquila, Italy
e-mail: francesca.micheli@graduate.univaq.it

L. Rossi
Department of Physical and Chemical Sciences,
University of L'Aquila, 67100 L'Aquila, Italy

© Springer Fachmedien Wiesbaden 2015
G. Dell, C. Egger (eds.), *World sustainable energy days next 2014*,
DOI 10.1007/978-3-658-04355-1_18

Since 1868, the concept of combining reactive and separation processes has been known [2], and nowadays different plant configurations could be designed to maximize H_2 yield in biomass gasification coupled with CO_2 capture, known as sorption-enhanced reaction processes (SERP).

In this work, the same synthesis method has been used to prepare CO_2 sorbents for high- and low-temperature processes to remove CO_2 directly into the gasifier and during the water gas shift (SEWGS) or hydrocarbons steam reforming (SER) reactions: improving the efficiency of the CO_2 capture processes maximizes hydrogen production. The materials used are much cheaper than the "state-of-the-art solution" proposed for precombustion capture, such as amines and Pd membranes; moreover, through process integration also the equipment volume decreases with a direct impact on the investment costs [3].

An interesting inorganic compound among the anionic clay family is hydrotalcite. This mineral is generally used after being thermally treated and the derived mixed oxides have interesting properties such as high superficial area, basic properties, formation of homogeneous mixtures of oxides with very small crystal sizes, and memory effect [4]. Thermally treated hydrotalcite could be used as a precursor, catalyst, and CO_2 sorbent. Hydrotalcite has a brucite-like structure, with M^{2+} cations octahedrally coordinated to hydroxyl groups; bivalent cations are partially substituted by trivalent ones, this increases the positive charge that is balanced by interlayer anions.

It is known that K_2CO_3 increases the basicity of sorbent surfaces enhancing CO_2 capture [5], and it is as well known that K_2CO_3 during the gasification process reduces coke formation on the catalyst surface, increasing its stability [6]. K_2CO_3 and $KHCO_3$ could increase hydrogen yield, as the alkali salts contained in biomass [7].

18.2 Materials and Method

Three kinds of sorbents were synthesized using low supersaturation method as shown in Table 18.1. At first, a hydrotalcite with nominal atom ratio Mg:Al=2, second, maintaining the ratio between $M^{2+}/M^{3+}=2/1$ constant, a "hydrocalumite type" [8] sorbent Ca/Al, and third, a Mg/Ca/Al mixed structure compound.

Table 18.1 Name and description of different hydrotalcites

Compound acronym	Description
H1	Hydrotalcites based on Mg and Al
H1K	Hydrotalcites based on Mg and Al+20% K_2CO_3
HT2	Hydrocalumite type based on Ca and Al
HT2K	Hydrocalumite type based on Ca and Al+20% K_2CO_3
HT3	Hydrotalcites/hydrocalumite type based on Mg, Ca, and Al
HT3K	Hydrotalcites/hydrocalumite type based on Mg, Ca, and Al+20% K_2CO_3

After synthesis, the compounds have been dried for 24 h at 120 °C and thermally treated for 8 h at 700 °C with a temperature ramp of 10 °C/min; after thermal treatment, they have been crushed and sieved within a particle diameter range of 355–710 μm.

The sieved fraction has been wet impregnated with 20 %$_{w/w}$ K$_2$CO$_3$ then dried and thermally treated again with the same temperature program.

For simplicity, every sorbent has been named with letters corresponding to H = hydrotalcite/hydrocalumite, T = type, K = potassium carbonate impregnated.

Specific surface area, pore size, and pore volume of samples were characterized by Brunauer–Emmett–Teller (BET) and Barrett–Joyner–Halenda (BJH) analysis, crystal phases by X-ray diffraction (XRD), morphology, and composition by scanning electron microscopy (SEM)energy-dispersive X-ray spectroscopy (EDX).

High- and low-temperature capture tests have been carried out in a fluidized bed reactor and simultaneous thermogravimetric analysis (TGA) and differential thermal analysis (DTA) were used to investigate the sorption and desorption behavior of sorbents and their cycle stability.

Under N$_2$ flow, a first desorption is made with a temperature ramp of 10 °C/min, for 30 min; sorbents are maintained at the regeneration temperature ($T_{des-high} = 700$ °C; $T_{des-low} = 450$ °C), then a decreasing temperature ramp of 4 °C/min allows to reach the carbonation temperature ($T_{ads-high} = 600$ °C; $T_{ads-low} = 350$ °C) and pure CO$_2$ flows for 1 h over the sorbent sample.

The same temperature was fixed during tests in the fluidized bed, where a positive CO$_2$ step injection is made when adsorption starts. This step lasts until CO$_2$% in the exit stream reaches the inlet value, then the regeneration step is made under N$_2$ flow with a temperature ramp of 100 °C, without any isothermal step; the temperature is finally decreased to the carbonation reaction condition. Sorbent stability was studied for three cycles in TGA–DTA analysis and five cycles in the fluidized bed reactor.

The sorbent dynamic CO$_2$ load has been evaluated by a first-order with dead time model proposed by Di Felice et al. [9] for gas mixing. The comparison with a blank test is the key model feature that allows to relate the instantaneous CO$_2$ concentration in the gas stream exiting the reactor with its uptake on the sorbent material in the reactor bed itself. A sorbent amount of 2–5 g was loaded into the reactor bed and mixed with 400 g of Merck sea sand (cod. 107712).

18.3 Results

18.3.1 Brunauer–Emmett–Teller and Barrett–Joyner–Halenda Analysis

Brunauer–Emmett–Teller and Barrett–Joyner–Halenda analysis have been performed on thermally treated samples after synthesis: H1 has the highest pore volume, average pore diameter, and specific surface area. All the sorbents show values proportional to the

Table 18.2 Specific surface area (BET), pore volume, and average pore diameter (BJH) analysis of HT1, HT2, HT3

	HT1	HT2	HT3
Specific surface area (m²/g)	203	100	149
Pore volume (cm³/g)	1.31	0.41	0.85
Pore diameter (Å)	236	157	203

BET, Brunauer–Emmett–Teller; *BJH*, Barrett–Joyner–Halenda

amount of magnesium: the greater the Mg content, the higher are the values obtained for these morphological properties (Table 18.2).

These results underline the synthesis process goodness and suggest the possibility to use such sorbents as a reactive support for a catalytic phase to carry out in one sorbent/catalyst body sorption-enhanced reaction processes (SER-SEWGS).

18.3.2 X-Ray Diffraction Analysis and Sorption Tests

The enhancement generated by K_2CO_3 in sorption capacity is higher in H1 (Mg–Al) tested at low temperature, whereas its effect is not well understood at high temperature where HT2 and HT3 show better performances when impregnation has not been made. In Fig. 18.1, the results from TGA–DTA analysis are shown; for the latter samples, the sorption capacity is, respectively, 4.78 and 3.91 $mmol_{CO2}$/g of dry sorbent. Sorption capacity evaluated by carbonation tests in the fluidized bed reactor shows the same trend during cycles, with much lower CO_2 loading, because of the lower CO_2 partial pressure; here, HT3 shows the best results on cycle stability and CO_2 sorption capacity is equal to 1.7 $mmol_{CO2}$/$g_{sorbent}$ at high-temperature carbonation condition (Fig. 18.2).

HT2K and HT3K show smaller capture capacity, due to the formation of fairchildite: a double carbonate of potassium and calcium $K_2Ca(CO_3)_2$ formed after the impregnation step, as it is shown by XRD analysis in Figs. 18.3 and 18.4d. In both HT2 and HT3 XRD spectra after test, carbonates formation is visible (Figs. 18.3 and 18.4c), according to CaO carbonation reaction at this operating condition [10, 11]. The DTA peak signal confirms the highly exothermal chemical reaction (170 kJ/mol). Differently, H1 and H1K XRD spectra are mostly amorphous (not shown in the figure); they do not show any carbonate phase, suggesting a physical or chemical sorption phenomenon on the aluminum amorphous phase. Silicon oxide from fluidized bed sand is also visible.

During high-temperature fluidized bed and TGA cyclic carbonation tests, formation of $Ca_{12}Al_{14}O_{32}$ has been found in HT2K and HT3K. Mayenite [12] produced by the reaction of CaO with Al_2O_3 ($12CaO + 7Al_2O_3 \rightarrow Ca_{12}Al_{14}O_{32}$) decreases the amount of CaO available for the carbonation reaction, thus explaining the smaller sorption capacity of the impregnated samples with respect to the unpromoted ones.

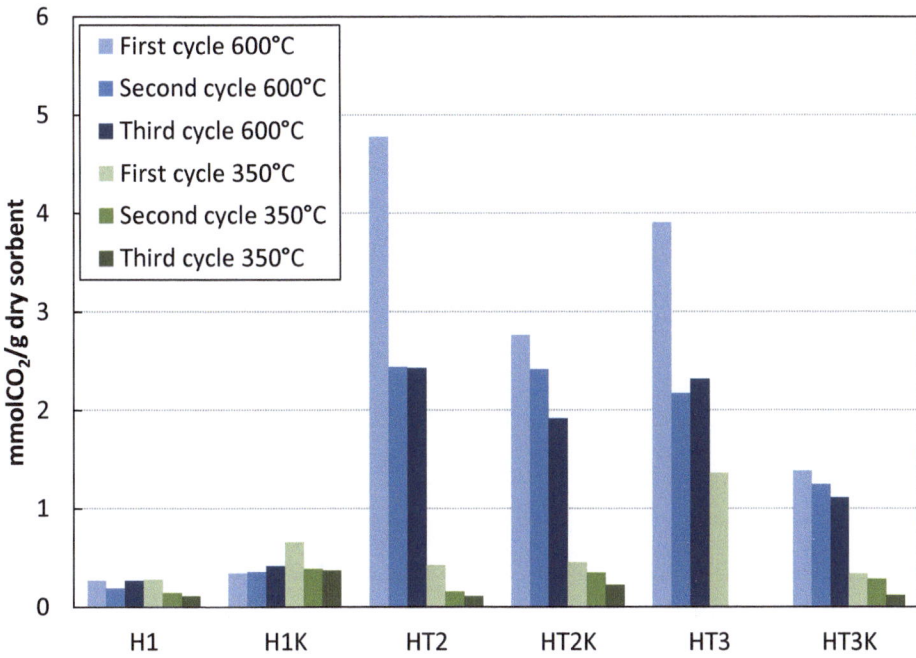

Fig. 18.1 Sorption capacity TGA–DTA analysis, P_{CO2} = 1 bar $T_{ads\text{-}low}$ = 350 °C, $T_{ads\text{-}high}$ = 600 °C

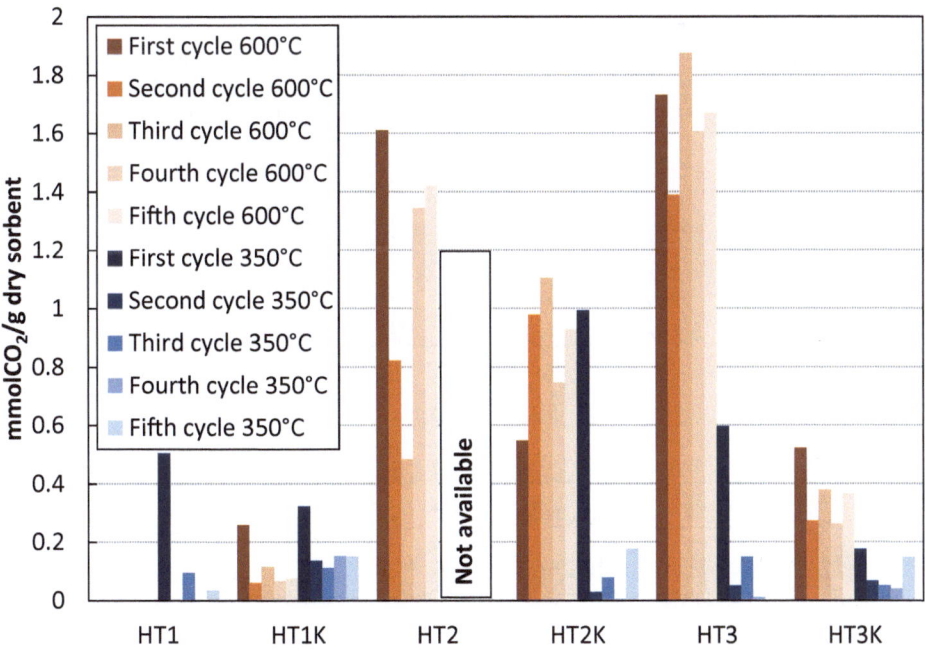

Fig. 18.2 Sorption capacity fluidized bed tests, P_{CO2} = 0.063 bar $T_{ads\text{-}low}$ = 350 °C, $T_{ads\text{-}high}$ = 600 °C

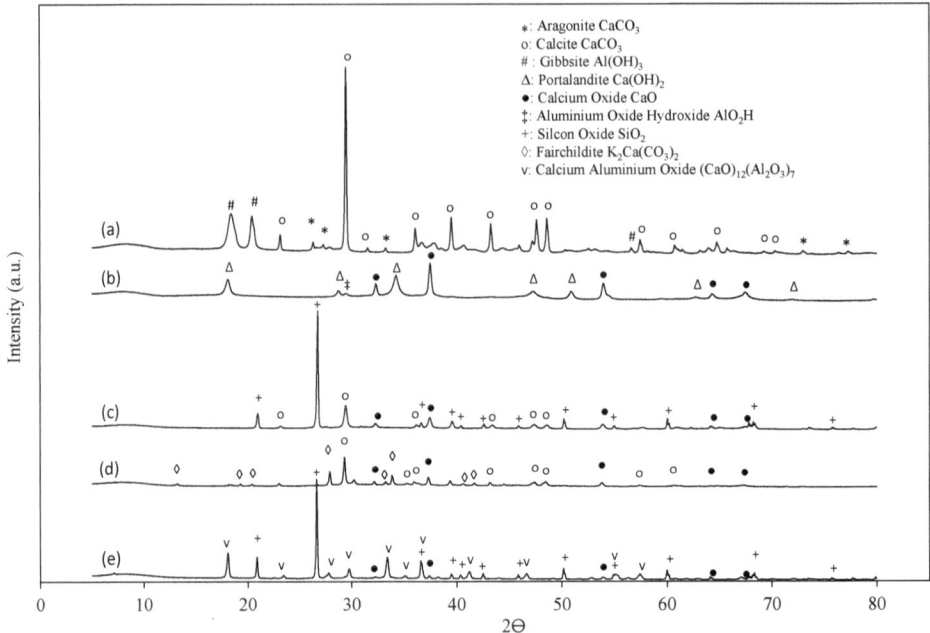

Fig. 18.3 XRD Pattern of **a** HT2 after drying at 120 °C for 24 h; **b** HT2 after thermal treatment at 700 °C for 8 h; **c** HT2 after capture cycles in fluidized bed at 600/700 °C; **d** HT2K after thermal treatment at 700 °C for 8 h; **e** HT2K after capture cycles in fluidized bed at 600/700 °C

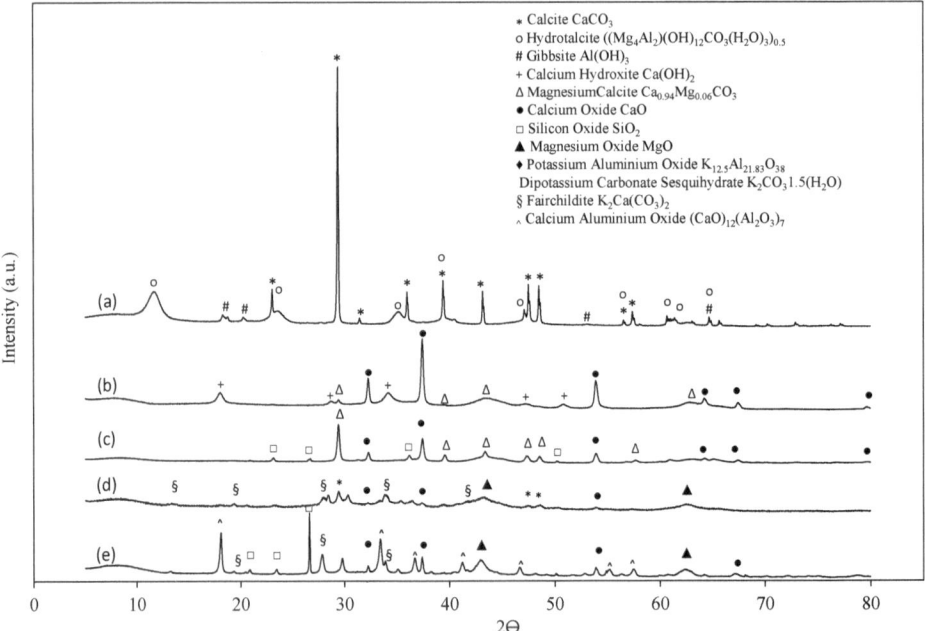

Fig. 18.4 XRD Pattern of **a** HT3 after drying at 120 °C for 24 h; **b** HT3 after thermal treatment at 700 °C for 8 h; **c** HT3 after capture cycles in fluidized bed at 600/700 °C; **d** HT3K after thermal treatment at 700 °C for 8 h; **e** HT3K after capture cycles in fluidized bed at 600/700 °C

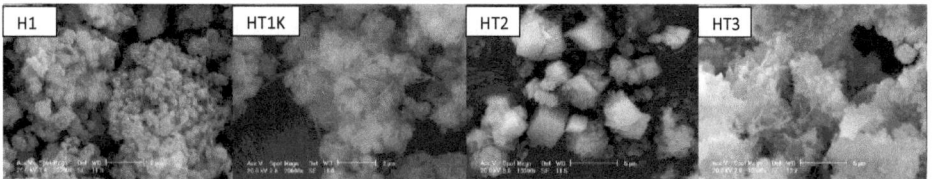

Fig. 18.5 SEM analysis of external surfaces of H1, H1K, HT2, HT3 after synthesis

Fig. 18.6 SEM analysis of external surfaces after fluidized bed reactor test: H1K tested at $T=350\,°C$ and $T=600\,°C$; HT2, HT3 tested at $T=600\,°C$

18.3.3   Scanning Electron Microscopy and Energy-Dispersive X-ray Analysis

To study and evaluate the morphology evolution of sorbents' external surface, SEM coupled with EDX analysis has been carried out on samples after synthesis, thermal treatment, and after low- and high-temperature fluidized bed reactor capture tests.

The characteristic "lamellae" shape of hydrotalcite is visible after synthesis in H1 and HT3 [13] in Fig. 18.5, while in HT2, cubic crystals of calcite are well visible [14]. These phases are also visible in XRD analysis spectra where hydrotalcite phases in H1 and HT3 and CaCO$_3$ rhombohedral structure of calcite in HT2 are diffracted (Figs. 18.3 and 18.4).

After K$_2$CO$_3$ impregnation, some needle-like structures are formed in H1K that disappear after calcination. HT2K and HT3K surfaces do not show substantial differences compared to the unpromoted surface (see Fig. 18.5).

Differently, the growth of needle-like shape structures on H1K is visible after tests. These structures seem to grow up during high-temperature tests: an external shell on the surface is formed; from EDX analysis, it results rich in Mg and probably has been promoted by the presence of potassium. HT2 and HT3 after high-temperature fluidized bed tests show calcium-rich agglomerates on the surfaces; in HT3, these agglomerates are dispersed in a smoother Mg–Al-rich matrix (see Fig. 18.6).

18.4   Conclusions

In this study, sorbents based on Mg–Al (H1), Ca–Al (HT2), and Mg–Ca–Al (HT3), with and without K$_2$CO$_3$ impregnation, have been synthesized, paying particular attention to the pH control during the synthesis and maturation of sorbents.

From the test results and characterization analysis, sorbents based on Mg–Al (H1–HT1K) work better at low temperature: Their sorption capacity at low temperature is higher than at high temperature. On the contrary, sorbents based on Ca–Al (HT2) and Mg–Ca–Al (HT3) work better at high temperature. Comparatively, the capacity of H1 at low temperature is much less than the capacity of HT2 and HT3 at high temperature and these results could be explained by the different nature of sorption phenomenon occurring: with H1 could be a physical adsorption, whereas HT2 and HT3 chemically adsorb CO_2, as it is clearly demonstrated by X-ray $CaCO_3$ phase detection and DTA peak signal.

It is found that K_2CO_3 impregnation has a positive effect only on sorbent based on Mg–Al (H1). The capacity of HT1K is a little bit higher than that of H1 at both low and high temperature, but still much less than the capacity of HT2 and HT3 at high temperature. The capacity of HT2 and HT3 at both low and high temperatures is higher than that of HT2 and HT3K. The low capacity of HT2K with respect to HT2 at low temperature is due to potassium and calcium carbonate mixed phases formation and calcium aluminate that decreases the availability of the sorbent (CaO) phase.

At high temperature, all sorbents based on Ca–Al (HT2) before and after impregnation, and sorbents based on Ca–Mg–Al (HT3) before and after impregnation, show good cyclic stability whereas at low temperature, they show a fairly good sorption capacity at first cycle that decreased or vanished in the following cycles because the regeneration temperature is too low to allow the calcination reaction, according to CaO carbonation thermodynamic equilibrium.

The high specific surfaces of these sorbents suggest them to be used as a reactive support for catalytic phases, to carry out in one sorbent/catalyst body sorption-enhanced reaction processes (SER-SEWGS).

Acknowledgments The authors kindly acknowledge the FCH JU UNIfHY for the financial support to this research project (contract 299732) and ASCENT FP7 (contract 608512).

References

1. UNIQUE (n.d.) *Seventh framework programme theme energy (G. A. 211517)* [Online] Available from: http://www.uniqueproject.eu [Accessed: 6 June 2014].
2. ROSTRUP-NIELSEN, J.R. (1984) Catalytic steam reforming, In J.R. Anderson, & M. Boudant (Eds)., *Catalysis science and technology*, 4, 1–118.
3. ASCENT (n.d.) *7FP (G.A. 608512) Advanced Solid Cycles with Efficient Novel Technologies.* [online] Available from: http://www.ascentproject.eu [Accessed: 9 June 2014].
4. CAVANI, F., TRIFIRÒ, F., VACCARI, A. (1991) Hydrotalcite-type anionic clays: preparation, properties and applications catalysis today. *Elsevier Science Publishers b.v*, II. 173–301.
5. LEE, J.M. et al. (2010) Enhancement of CO_2 sorption uptake on hydrotalcite by impregnation with K_2CO_3. *Langmuir*. 261. 8788–97.
6. KUCHONTHARA, P. et al. (2012) Catalytic steam reforming of biomass-derived tar for hydrogen production with $K_2CO_3/NiO/\gamma$-Al_2O_3 catalyst. *Korean J. Chem. Eng.* 29 (11). 1525–1530.

7. KRUSE, A., FAQUIR, M. (2007) Hydrothermal Biomass Gasification – Effects of Salts, Backmixing, and Their Interaction. *Chem. Eng. Technol.* 30 (6) 749–754

8. LOPEZ-SALINAS, E. et al. (1996) Characterization of Synthetic Hydrocalumite-Type [Ca$_2$Al(OH)$_6$]NO$_3$.mH$_2$0: Effect of the Calcination Temperature. *J. Porous Mat.* 2. 291–297.

9. DI FELICE, L., FOSCOLO, P.U., GIBILARO, L. (2011) CO$_2$ Capture by Calcined Dolomite in a Fluidized Bed: Experimental Data and Numerical Simulations. *Int. J. Chem. Reactor Eng.* 9: Article A. 55.

10. STANMORE, B.R., GILOT, P. (2005) Review—calcination and carbonation of limestone during thermal cycling for CO$_2$ sequestration. *Fuel Process. Technol.* 86. 1707–1743.

11. SILCOX, D., KRAMLICH, J.C., PERSHLING, D.W. (1989) A mathematical model for flash calcination of dispersed CaCO$_3$ and Ca(OH)$_2$ particles. *Ind. Eng. Chem. Res.* 28. 155–160.

12. PALACIOS, L. et al. (2007) Crystal structures and in-situ formation study of mayenite electrides. *Inorg. Chem.* 46, 4167–76.

13. WALSPURGER, S. et al. (2010) In Situ XRD Detection of Reversible Dawsonite Formation on Alkali Promoted Alumina: A Cheap Sorbent for CO$_2$ Capture. *Eur. J. Inorg. Chem.* 2461–2464.

14. SONDIA, I. et al. (2009) The electrokinetic properties of carbonates in aqueous media revisited Colloids and Surfaces A: Physicochem. Eng. Aspects. 342, 84–91.

The Effects of Torrefaction Parameters on the Thermochemical Properties of *Jatropha curcas* Seed Cake

19

Buddhike Neminda Madanayake, Carol Eastwick,
Suyin Gan and Hoon Kiat Ng

Abstract

The efficacy of torrefaction on *Jatropha curcas* seed cake as a viable candidate for co-firing with coal was investigated. Increasing the torrefaction temperature and holding time resulted in decreases in both mass and energy yields and an increase in higher heating value (HHV). The optimum torrefaction conditions were a temperature of 250 °C and a holding time below 30 min, which resulted in an increase in the HHV by 15.4–19.0 %, while the energy yield remained in the range of 93.67–94.51 %.

19.1 Introduction

The overdependence on fossil fuels as a source of energy is widely acknowledged as a growing cause for concern on a global scale. Amongst the many alternative energy sources that are currently available, biomass is a promising source as a co-firing feedstock. In co-firing, it is desirable to bring the physical characteristics of the biomass such as moisture

B. N. Madanayake (✉) · C. Eastwick
Energy and Sustainability Research Division, The University of Nottingham,
University Park, Nottingham NG7 2RD, UK
e-mail: enxbnm@nottingham.ac.uk

S. Gan
Department of Chemical and Environmental Engineering,
The University of Nottingham Malaysia Campus,
Jalan Broga, 43500 Semenyih, Selangor, Malaysia

H. K. Ng
Department of Mechanical, Materials and Manufacturing Engineering, The University of
Nottingham Malaysia Campus, Jalan Broga, 43500 Semenyih, Selangor, Malaysia

© Springer Fachmedien Wiesbaden 2015
G. Dell, C. Egger (eds.), *World sustainable energy days next 2014*,
DOI 10.1007/978-3-658-04355-1_19

content and energy density to as close as possible to that of coal. This would allow the biomass to be co-fired with coal in existing coal power plants with minimal modifications to the handling and combustion equipment [1]. Torrefaction is a thermal pretreatment technique that has shown promising results in accomplishing this. The process involves heating the biomass to a temperature in excess of 200 °C in a nitrogen (N_2) atmosphere, which results in a mild pyrolysis reaction with no combustion [2]. This has been shown to increase the calorific value and energy density and reduce the propensity to absorb moisture [3–5].

The feedstock of interest in this project is *Jatropha curcas*. *J. curcas* is garnering increasing attention as a source of bioenergy. It is resistant to droughts, and thrives in the arid conditions of Central and South America, Southeast Asia, India and Africa [6]. Another important characteristic of *Jatropha* is its toxicity (which makes it nonedible)—this ensures that the energy industry does not have to compete with the food industry over the plant's bioresources, as is the case with other bioenergy sources such as palm oil and sugar cane [7]. This in turn improves the economic viability of using the products from *Jatropha*. The oil extracted from *J. curcas* seeds has been shown to be a viable feedstock for biodiesel production [8]. However, the yield of oil is only about 18 % by mass of the dry fruit [9]. A considerable amount of solid biomass is discarded as waste, and there is the potential to extract energy from this waste seed cake as well. Investigating the feasibility of combusting the *Jatropha* shells and seed husks is thus an important avenue of research in this field. However, like all biomass, the co-firing of *Jatropha* seed cake could potentially prove to be problematic due to the differences in properties with coal. Thus, it is useful to explore pretreatment techniques such as those mentioned previously. In the light of this, this project aims to investigate the effects of torrefaction on certain thermochemical properties of *J. curcas* seed cake.

19.2 Methodology

19.2.1 Overview

The fundamental objective of the experimental work carried out was to investigate the effects of torrefaction parameters, i.e. torrefaction temperature and holding time on the quality of the solid torrefied product. A wide temperature range was chosen for the current investigation since it was a preliminary one to establish the overall picture. Three temperatures (200, 250, 300 °C), and three holding times (0, 30, 60 min), were chosen. The '0 min' holding time was included in the experimental matrix to gauge the degree of torrefaction occurring during the heating and cooling segments of the run. Each temperature–time combination was carried out in triplicate. Hence, a total of 27 torrefaction runs were carried out. Measurements of the mass of the sample before and after the run enabled calculation of the mass yield. The torrefied samples were then subjected to two analyses. A bomb calorimeter (PARR Instruments 6100) was used to measure the higher heating value or HHV (which subsequently enables the calculation of the energy yield). A thermogravimetric analysis (TGA) was carried out to determine the proximate composition of the torrefied biomass, i.e. moisture content, volatile matter content, fixed carbon content and ash content.

19.2.2   TGA

The TGA unit used was a Mettler Toledo TGA/DSC 1. Based on the conditions outlined in British Standards' publications pertaining to the proximate analysis of biomass (BS EN 14774, BS EN 15148, BS EN 14775), the following TGA profile was devised:

- Under N_2 flowing at 50 mL/min
  - Heating to 105 °C at 10 °C/min; holding for 40 min
  - Heating to 905 °C at 20 °C/min; holding for 7 min
  - Cooling to 550 °C at 20 °C/min
- Under O_2 flowing at 50 mL/min
  - Holding at 550 °C for 120 min

19.2.3   Torrefaction

The torrefaction runs were carried out in a tubular furnace (Carbolite model CTF 12/65/550) with an attached rotameter to control the gas flow rate. For each torrefaction run, a measured quantity of ground *Jatropha* seed cake (approximately 1.5 g) was placed in a ceramic 'weighing boat' inside the furnace. An N_2 flow of 100 mL/min at room temperature was used. A proportional-integral-derivative (PID) controller was used to programme the temperature ramp and holding time. After the run was over, the furnace was allowed to cool to 150 °C before the sample was removed and allowed to cool further in a dessicator. When the sample had cooled down to room temperature, the mass was measured again. Following weighing, each sample was transferred to an airtight glass vial for storage until subsequent analyses were carried out.

19.3   Results and Discussion

19.3.1   Mass Yield, HHV, Energy Yield

The HHV of the untreated *Jatropha* was found to be 17.682 MJ/kg. This was a mean value calculated from five calorimeter runs. This mean value was used to calculate the energy yields following torrefaction. The mass and energy yields were calculated using the following Eqs. 19.1 and 19.2, respectively:

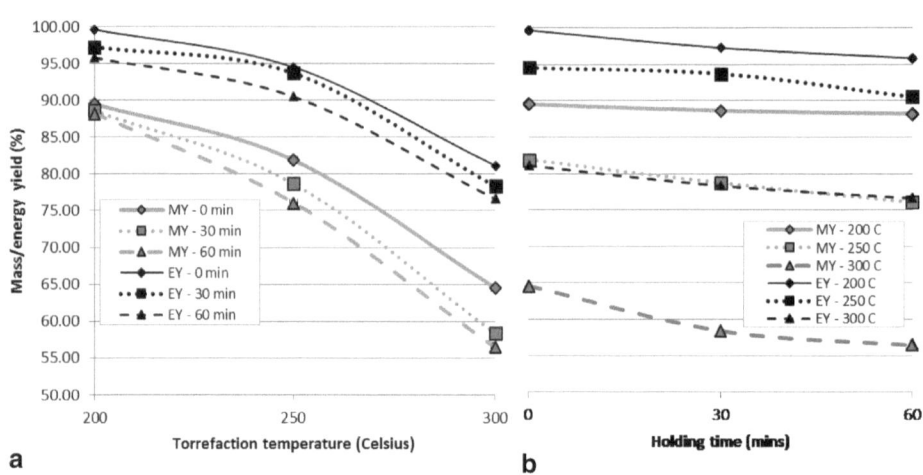

Fig. 19.1 Variation of mass yield (*MY*) and energy yield (*EY*) with **a** temperature and **b** time

$$mass\ yield = \frac{final\ mass}{initial\ mass} \times 100\% \qquad (19.1)$$

$$energy\ yield = \frac{HHV\ of\ torrefied\ sample \times final\ mass}{HHV\ of\ raw\ sample \times initial\ mass} \times 100\% \qquad (19.2)$$

Figure 19.1 illustrates the effect of torrefaction temperature and holding time on the mass yield. It can be observed that there is a reduction in the mass yield with increasing torrefaction temperature and holding time. This is in compliance with the expected results, since torrefaction involves the loss of both moisture and volatile matter from the sample of biomass. However, it can be seen when comparing the two charts that the increase in torrefaction temperature had a much larger effect on the mass yield than the increase in holding time. When the torrefaction temperature is increased from 200 to 300 °C, there is a decrease of 24.93, 30.17 and 31.67 % for 0, 30 and 60 min holding times, respectively. When the holding time is increased from 0 to 60 min, there is a decrease of 1.32, 5.81 and 8.06 % for torrefaction temperatures of 200, 250 and 300 °C, respectively. The decrease in mass yield due to an increase in the torrefaction temperature is thus about an order of magnitude higher than that due to an increase in the holding time.

Figure 19.2 illustrates the effect of torrefaction temperature and holding time on the HHV. An increase in torrefaction temperature clearly leads to a substantial increase in the HHV. Increasing the torrefaction temperature from 200 to 300 °C resulted in an increase of the HHV by 2.541, 4.307 and 4.772 MJ/kg for the 0, 30 and 60 min holding times, respectively. In contrast, the effect of holding time on HHV; however, was much less significant. Increasing the holding time from 0 to 60 min caused an increase in HHV by 0.63

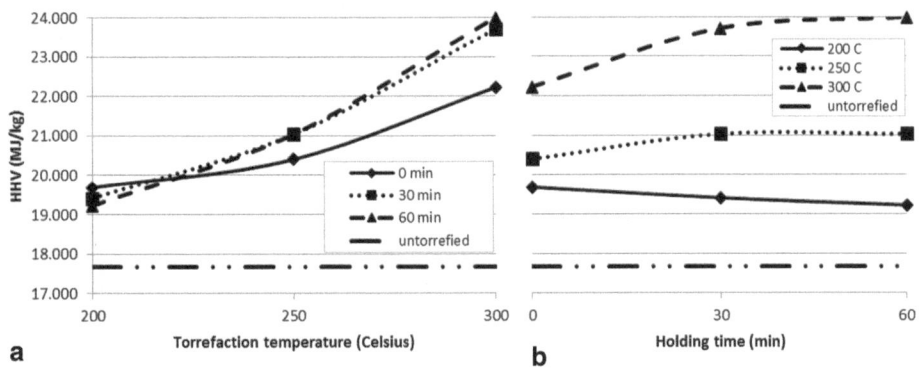

Fig. 19.2 Variation of higher heating value (*HHV*) with **a** temperature and **b** time

and 1.765 MJ/kg for torrefaction temperatures of 250 and 300 °C, respectively, while the HHV decreased by 0.466 MJ/kg for the 200 °C case.

After torrefaction, the HHV has increased from the untreated value of 17.682 MJ/kg by a minimum of 8.7 %. The most intense torrefaction conditions (300 °C for 60 min) resulted in an increase by 35.7 %. This increase has been explained in literature by the fact that torrefaction results in a loss of both moisture and volatile matter. Loss of moisture results in a reduction of hydrogen (H) and oxygen (O) atoms from the biomass without a loss of carbon (C). Volatile matter lost from biomass consists of hydrogen and light hydrocarbons which have a low C content compared to H [10]. Hence, there is a substantially greater amount of H and O lost during torrefaction than there is C [2]. There is a reduction in the H/C and O/C ratios, and since the C–H and C–O bond energies are lower than that of C–C, this results in an increase in the calorific value of the biomass [3, 11].

The effect of torrefaction temperature and holding time on the energy yield of the torrefied product is also shown in Fig. 19.1. Again, the effect of varying the torrefaction temperature is more prominent than that of changing the holding time. Increasing the holding time from 0 to 60 min caused a decrease in the energy yield of 3.79, 4.00 and 4.49 % for torrefaction temperatures of 200, 250 and 300 °C, respectively. Increasing the torrefaction temperature from 200 to 300 °C caused more significant decreases of 18.49, 18.90 and 19.19 %, for holding times of 0, 30 and 60 min, respectively.

The energy yield is a function of the mass yield and the HHV. With increasing torrefaction intensity, the HHV increases and this would have the effect of increasing the energy yield. However, the mass yield decreases with increasing torrefaction intensity, and this has a negative effect on the energy yield. The overall decrease in energy yield is due to the fact that the increase in HHV is not sufficient to compensate for the reduction in the mass yield—torrefaction has a more pronounced effect on the mass yield than on the HHV.

A feature common to the mass yield, HHV and energy yield is that all three are affected much more significantly by a change in torrefaction temperature than by a change in holding time. A linear increase in the torrefaction temperature resulted in an almost

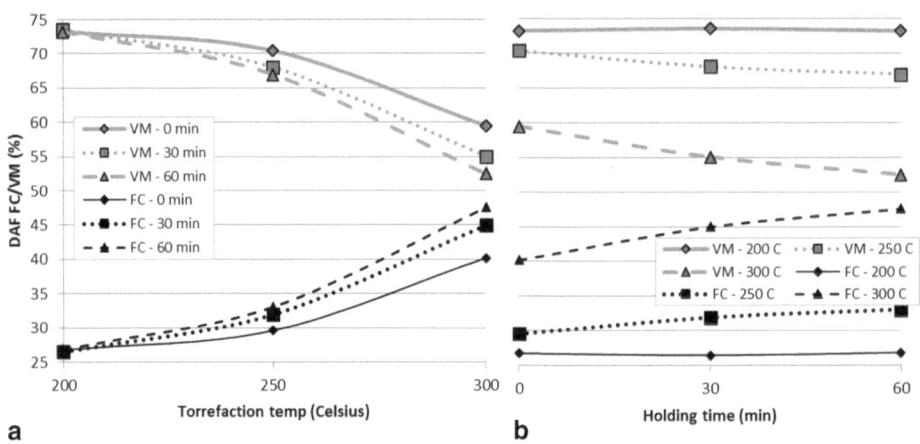

Fig. 19.3 Variation of DAF FC and DAF VM with **a** temperature and **b** time

exponential decrease and increase in the mass yield and HHV, respectively. Thus, when determining the optimal torrefaction conditions for *Jatropha* seed cake, determining the ideal torrefaction temperature is of paramount importance.

19.3.2 Fixed Carbon (FC) and Volatile Matter (VM) Content

Figure 19.3 shows the variation of the dry ash-free (DAF) FC and VM content in the torrefied samples. The variation of the DAF FC and DAF VM are inversely proportional to each other, as expected. As in the cases of mass yield and HHV, the DAF FC and VM, both show a stronger response to variations in torrefaction temperature than to holding time. When the holding time was increased from 0 to 60 min, there was an absolute increase of the DAF FC of 0.07 (percentage by weight), 3.46 and 7.31% for torrefaction temperatures of 200, 250 and 300 °C, respectively. An increase in the torrefaction temperature from 200 to 300 °C caused a rise of 13.43, 18.48 and 20.67% for holding times of 0, 30 and 60 min, respectively.

The response of the VM content to higher torrefaction temperatures could be explained in terms of the decomposition mechanisms. Greater loss of VM when higher temperatures are utilised could be attributed to the initiation of certain decomposition pathways which require higher activation energies. Since VM consists mostly of low-C molecules, a lower fraction of C is lost from the biomass compared to other elements such as H. This would inevitably lead to an increase in the FC content. This would subsequently lead to a reduction in H/C and O/C ratios and explain the increase in the HHV which was reported earlier.

19.3.3   Discussion

Torrefaction temperature was found to be the dominant factor (over the holding time) in affecting the properties of the torrefied samples. It was observed that when the torrefaction temperature was increased linearly, there was a near-exponential change in the properties (that were investigated) of the torrefied samples. The mass yield and energy yield decreased. The fixed carbon content and HHV increased. A higher HHV is a desirable trait, since it translates to a higher energy density. This in turn means that a smaller amount of the treated biomass is required to yield a certain amount of energy, and this is advantageous in terms of transport, handling and storage. However, the higher torrefaction temperatures which are needed to increase the HHV also lead to a reduction in the mass and energy yields. A low energy yield means that a larger amount of the untreated biomass is required to produce a certain amount of energy. This is undesirable in terms of economics.

Hence, it is important to consider the trade-off between a higher calorific value and lower energy yield, and decide on a torrefaction temperature which gives the best compromise. Torrefaction at 200 °C, although resulting in a high energy yield (between 95.8 and 99.6 %), did not cause an increase in the HHV which was substantial enough to make torrefaction worthwhile. On the other hand, torrefaction at 300 °C caused the energy yield to drop to between 76.7 and 81.1 %—the loss in the total energy content was considered too low in this case. The intermediate temperature of 250 °C appeared to provide a compromise, as it caused an increase in the HHV by 15.4–19.0 %, while maintaining the energy yield at acceptable levels (between 90.5 and 94.5 %).

With respect to a linear increase in holding time, the change in properties was not as marked as those observed with an increase in torrefaction temperature. In most cases, there was a linear but not-negligible change in the properties. Taking into consideration the aforementioned factors involving HHV and energy yield, as well as the energy costs associated with keeping the sample at an elevated temperature for an extended period of time, a holding time between 0 and 30 min appears to be ideal.

19.4   Conclusions

As expected due to the loss of volatiles, the mass yield was found to decrease with increasing torrefaction intensity, i.e. increasing torrefaction temperature and holding time. Although the HHV increased, it could not compensate for the reduction in mass yield, and thus there was a decrease in the energy yield as well. TGA of the torrefied products confirmed the reduction in the VM content as the torrefaction was intensified. Conversely, there was an increase in the FC content. Although these general trends were observed with both increasing torrefaction temperature and increasing holding time, the temperature had a more significant effect on the characteristics of the torrefied product than the holding time did. Based on the results obtained, the optimum temperature was determined to be 250 °C, while the optimum holding time was less than 30 min. These conditions resulted in the HHV increasing by 15.4–19.0%, while maintaining the energy yield between 93.67 and 94.51%. The parameters chosen for this study encompassed a wide range in order to establish preliminary guidelines. Hence, further investigations should be carried out in order to optimise the recommended torrefaction conditions to a finer degree.

References

1. HUGHES, E., TILLMAN, D. (1998) Biomass cofiring: status and prospects 1996. *Fuel processing technology,* 54(1–3), 127–142.
2. CHEN, W., KUO, P.-C. (2010) A study on torrefaction of various biomass materials and its impact on lignocellulosic structure simulated by a thermogravimetry. *Energy*, 35(6), 2580–2586.
3. McKENDRY, P. (2002) Energy production from biomass (Part 1): Overview of biomass. *Bioresource technology*, 83(1), 37–46.
4. PIMCHUAI, A., DUTTA, A., BASU, P. (2010) Torrefaction of Agriculture Residue To Enhance Combustible Properties. *Energy & Fuels*, 24(9), 4638–4645.
5. SADAKA, S., NEGI, S. (2009) Improvements of biomass physical and thermochemical characteristics via torrefaction process. *Environmental progress & sustainable* ..., 28(3), 427–434.
6. GÜBITZ, G., MITTELBACH, M., TRABI, M. (1999) Exploitation of the tropical oil seed plant *Jatropha curcas* L. *Bioresource Technology*, 67, 73–82.
7. WEVER, D.-A.Z., HEERES, H.J., BROEKHUIS, A.a. (2012) Characterization of Physic nut (*Jatropha curcas* L.) shells. *Biomass and Bioenergy*, 37, 177–187.
8. SRICHAROENCHAIKUL, V., ATONG, D. (2009) Thermal decomposition study on Jatropha curcas L. waste using TGA and fixed bed reactor. *Journal of Analytical and Applied Pyrolysis*, 85(1–2), 155–162.
9. SINGH, R.N. et al. (2008) SPRERI experience on holistic approach to utilize all parts of *Jatropha curcas* fruit for energy. *Renewable Energy*, 33(8), 1868–1873.
10. VASSILEV, S. V. (2010) An overview of the chemical composition of biomass. *Fuel*, 89(5), 913–933.
11. NARVAEZ, I., ORIO, A., 1996. Biomass gasification with air in an atmospheric bubbling fluidized bed. Effect of six operational variables on the quality of the produced raw gas. *Industrial & Engineering* ..., 5885(95), 2110–2120.

REVE: Versatile Continuous Pre/Post-Torrefaction Unit for Pellets Production

20

Nicolas Doassans-Carrère, Sébastien Muller and Martin Mitzkat

Abstract

Torrefaction is a well-known biomass pretreatment for energy densification and preservation enhancement, but continuous industrial plants are still rare. Pelletization, a well-known energy densification process as well, can be coupled with torrefaction in order to increase both energy density and water-resistivity of pellets. This chapter presents the use of a proven innovative technology coming from the food industry for biomass torrefaction before (pretorrefaction) and after (post-torrefaction) the pelletization step. Several biomasses (wood chips, coniferous barks, olive pips, straw, and pine pellets) have been torrefied and analyzed. The choice between pre- and post-torrefaction is finally discussed.

20.1 Introduction

Biomass is generally admitted as a great potential renewable source for heat and power, regarded for the increase of global energy demand, the depletion of fossil fuels, and the regulation of CO_2 emission. However, some efforts in handling, conversion, and transport of biomass remain to be done [1].

Several thermochemical conversion processes have been studied: fast pyrolysis and direct liquefaction to produce bio-oils [2], gasification to produce syngas [3], and slow pyrolysis (torrefaction) or hydrothermal carbonization to produce solid bio-fuels [4]. The diversity of biomass composition and properties makes their use complicated and a pre-

N. Doassans-Carrère (✉) · S. Muller · M. Mitzkat
Loriol-sur-Drôme, France
e-mail: nicolas.doassans@revtech.fr

© Springer Fachmedien Wiesbaden 2015
G. Dell, C. Egger (eds.), *World sustainable energy days next 2014*,
DOI 10.1007/978-3-658-04355-1_20

treatment is often necessary. Torrefaction also appears as the most promising pretreatment to increase the uniformity of feedstock, the energy density, and some properties like hydrophobicity or grindability. Typically, with a reactor temperature ranging from 200 to 300 °C for 10 min to 2 h, in the absence or low concentration of oxygen, a solid mass yield of 70 % is achieved from woody biomass, containing 90 % of the initial energy content [5].

Pelletization, a mechanical biomass densification process, is currently the only supply chain of solid biofuels (pellets) globally developed [6]. The biomass densification in pellet mill (from 0.2–0.3 g/cm³ to 0.65–0.75 g/cm³) makes pellets good for transport costs and handling [7]. However, the moisture uptake ability of pellets prevents long bulk storage and transport [8]. Thus, the combination of pelletization and torrefaction (before or after the pelletization step) has been studied and appears as an effective way to produce high energy density and water-resistant pellets [6, 8, 9].

Laboratory-scale torrefaction reactors are common but continuous industrial torrefaction plants are still rare. Technical challenges like gas handling, process upscaling, product quality, flexibility in using different input material, or heat integration have to be met [10].

The present chapter presents a robust, simple, and modern technology coming from the food industry. Pre-torrefaction is studied with four biomasses (wood chips, coniferous barks, olive pips, and straw) as well as post-torrefaction with pine pellets. The integration of this torrefaction unit in an existing pelletizing plant is then discussed.

20.2   Materials and Methods

20.2.1   Technology and Pilot Unit

Vibrating electrical elevator and reactor (REVE) is based on three simple principles: transport and mixing in a stainless steel spiral tube by vibrations; heating by electricity flowing in the wall of the tube; and atmosphere control in the confined spiral tube [11]. Such industrial machines have been successfully installed for roasting various food ingredients such as peanuts, barley, or soya with flowrates up to 5 ton/h.

Figure 20.1a is an example of a 300 kg/h REVE (depending on product density), composed by a torrefaction part (at the bottom, insulated) and a cooling part (at the top). Inside the whole REVE, the atmosphere is controlled, preventing fire during torrefaction and cooling. At the end of the cooling part, if more temperature drop is required (storage for example) a cooling tunnel can be used, without any risk of fire due to the low inlet product temperature.

Experimental studies were carried out using a pilot unit of the Revtech Research & Development department (Fig. 20.1b), with the following characteristics: a spiral tube with an external diameter of 88.9 mm and a total length of 33.3 m on eight turns, vapors and gases extraction at all turns of the spiral tube into the manifold, and a solids recirculation system in order to increase residence times and sample easily [11].

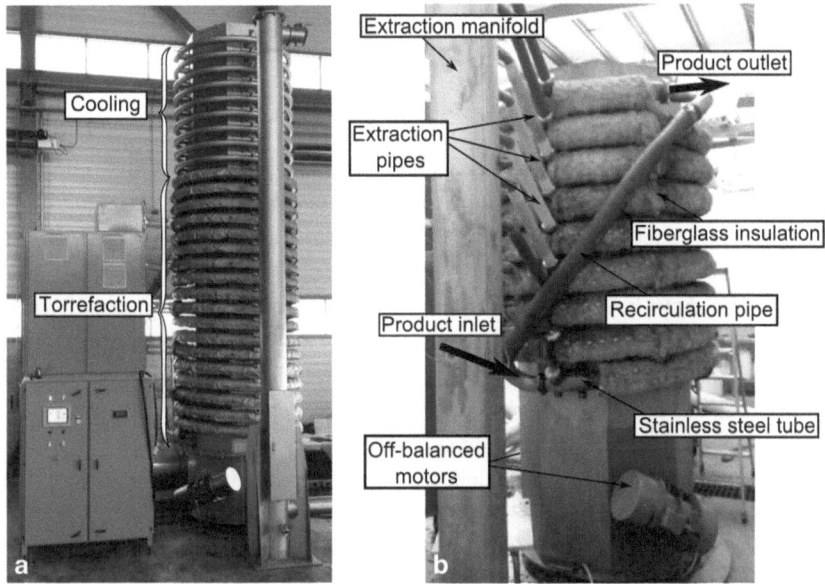

Fig. 20.1 Example of a 300-kg/h torrefaction-cooling unit (**a**), and REVE pilot unit (**b**)

Table 20.1 Feedstock properties

Biomass	Moisture content (%)	Density (g/cm³)	Particles size (mm)
Wood chips	11.8	0.28	5–20
Coniferous barks	59.7	0.33	10–40
Olive pips	20.3	0.66	5
Straw	11.0	0.05	20–100
Pellets din+	<8%	>0.65	6*30

20.2.2   Feedstock and Operating Conditions

Five biomasses (wood chips, coniferous barks, olive pips, and straw) have been selected for their different properties like density, moisture content, and particle size distributions (Table 20.1).

Depending on the biomass density and particle size, mass flowrates studied with the pilot unit are different from a biomass to another: 10 kg/h of straw, 85 kg/h of wood chips, 95 kg/h of coniferous barks, 140 kg/h of olive pipes, and 200 kg/h for pellets. Several tube temperatures have been studied with residence times up to 30 min: 240 °C (straw), 260–280 °C (wood chips, coniferous barks, olive pipes, and pellets), and 300 °C (wood chips).

20.3   Results and Discussion

20.3.1   Pre-Torrefaction

Pelletization requires small biomass particles (<3–4 mm) with a maximum inlet moisture content of 10–15 % [7]. This means the use in most cases of drying and grinding steps. Pre-torrefaction can be used as a pretreatment for pelletization in order to increase uniformity of feedstock (humidity, grindability).

All biomasses have been torrefied to obtain biochar from wood chips, coniferous barks and olive pips. Figure 20.2 shows some examples of color evolution, according to residence time.

For wood chips, coniferous barks, and olive pips, there is clearly a noncolor change period which corresponds to the heating and drying steps (ranging 10–15 min). Once this period finished, dark coloration is quick. In the straw case, this period is very short because of the low initial moisture and the thickness of the straw.

Torrefaction impacts both elemental composition and grindability of biomass. At the most severe conditions, biochars obtained from wood, olive pips, and coniferous barks are closed lignite [11] which can be useful for applications like gasification or as a coal substitute [5]. However, for pelletization pretreatment, lower torrefaction degrees would be more adapted (see Sect. 20.3.2).

Fig. 20.2 Coloration of torrefied products according to residence time

Table 20.2 Mass yields (a_r) and energy yields of torrefied wood chips at 85 kg/h [11]

T (°C)	t (min)	Y_S (wt%)	η_S (%)	d_η
260	30	74.2	89.4	1.20
280	30	64.3	80.4	1.25
300	15	59.4	77.7	1.31
300	30	50.8	68.6	1.35

Using a blade grinder, grindability improvement has been evaluated for each biomass. Wood chips and olive pips are respectively poorly grindable and not grindable with the blade grinder. However, biochars from these types of biomass become easily grindable. Torrefaction impact on grindability is lower for coniferous barks and straw which keep respectively easily and poorly grindable [11].

Table 20.2 gives some examples of solids mass yields, Y_S, obtained for the torrefaction of wood chips, defined on as received (a_r) basis. Using higher heating values (HHV), energy yields, η_S, can be estimated. From an initial HHV of 19.12 MJ/kg, energy yields range from 68.6% at 300 °C (HHV: 22.78 MJ/kg) to 89.4% at 260 °C (HHV: 20.31 MJ/kg). Energy densification (d_η) can also be estimated, which is equal to the energy yield divided by the mass yield.

With the increase of torrefaction degree, there is an increase of the energy densification, which means that the initial biomass energy is more and more concentrated in the biochar. However, this energy densification also goes with a high decrease of mass yield, which should be limited for pellets production. In the case of wood chips and olive pips, pre-torrefaction increased both the energy densification and the grindability, reducing the cost of the grinding step. Thus, pre-torrefaction for these kinds of biomass could be interesting. However, a low degree of torrefaction is more desirable in order to keep a good mass yield (70–80%) and to not reduce the pelletizing properties of materials [9].

On the other hand, the low torrefaction impact on grindability of coniferous barks and straw makes the pre-torrefaction for these kinds of biomass more disputable. Moreover, the high moisture content of coniferous barks may make a drying step more adapted than pre-torrefaction. The extremely low density of straw entails a low mass flowrate inside the REVE and increases the torrefaction cost per kilogram. In that case, post-torrefaction of straw pellets seems to be more adapted.

20.3.2 Post-Torrefaction

In some cases, for example with straw (see Sect. 20.3.1), it could be more desirable to do the torrefaction after the pelletization: post-torrefaction. Moreover, plenty of pellets production plants are already running and post-torrefaction could be more easily integrated. The main effects of post-torrefaction looked for are the energy densification and the water-resistivity.

Raw   260°C - 15' / 280°C - 9'   260°C - 27'   280°C - 15'

Fig. 20.3 Examples of torrefied pellets according to temperature and residence time

Post-torrefaction inside the REVE produces good quality torrefied pellets (Fig. 20.3). Pellets are not crushed or crumbled and torrefaction degree (color) is homogeneous in both core and shell.

Likewise for pre-torrefaction, really dark torrefied pellets can be obtained in short time (Fig. 20.3, 280 °C–15′). However, higher torrefaction degree means lower mass yield. Seeing that one of the main pellets properties looked for is the hydrophobicity, moisture uptake tests are performed.

Figure 20.4 shows the behavior of raw and torrefied pellets soaked into water, simulating a quick moisture uptake. The test reveals clearly that pellets torrefied above 15 min at 260 °C are hydrophobic and suitable for long bulk storage. This kind of pellets (d) is also obtained after 9 min at 280 °C (Fig. 20.3), which indicates that good hydrophobic torrefied pellets can be produced in short time into the REVE. Higher torrefaction degrees do not seem to be necessary and would only imply lower mass yields. Torrefied pellets (d) in Fig. 20.4 show torrefaction degree similar to wood chips torrefied at 260 °C (similar volumetric flowrate, moisture content, and torrefaction severity). Thus, it can be considered that they are produced with a mass yield of around 75 % and an energy yield of around 90 %. Post-torrefaction of pellets appears to be a good way to produce water resistant solid bio-fuel with a higher energy density.

20.3.3   Discussion

The REVE is a simple and compact unit (Fig. 20.1) which can be easily integrated in a new or existing pellets production plant, as a pre- or post-torrefaction step. The choice between pre-torrefaction and post-torrefaction depends on feedstock and unit integration. However, for both cases, a low degree of torrefaction is sufficient in order to have a good energy densification (without too much mass losses), an increase of grindability and/or a good hydrophobicity.

For an existing pellets production plant, post-torrefaction seems to be the best choice. Indeed, investments for every process step from raw biomass to pellets are already made and are functioning. The added value of torrefied pellets (energy density and hydrophobicity) should be reached within a simple unit capable to be easily integrated at the end of the production line. However, if the goal is the feedstock diversification, pre-torrefaction

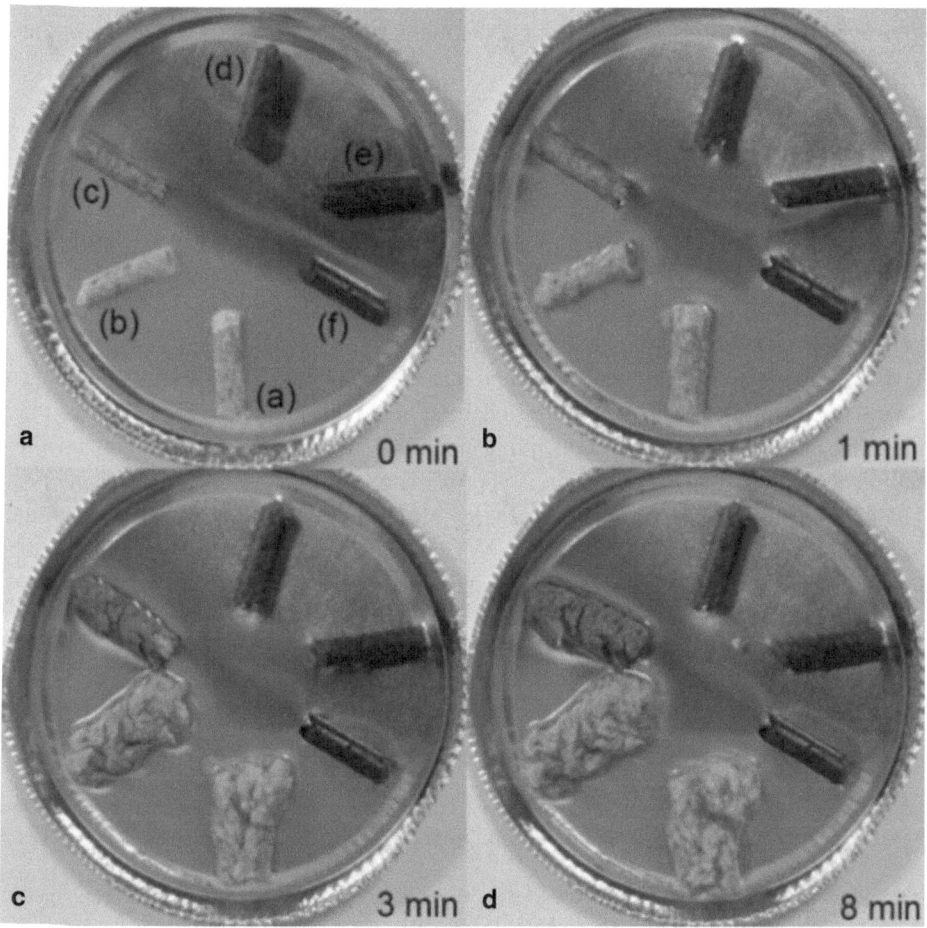

Fig. 20.4 Water soak of **a** raw pellet, and pellets torrefied at 260 °C for **b** 3 min, **c** 9 min, **d** 15 min, **e** 21 min, and **f** 27 min

could be interesting in order to increase the inlet material uniformity. For a new pellets production plants, the choice will be mainly based on the inlet biomass. For some biomass like wood chips or olive pips, the gain in grindability using pre-torrefaction could impact the cost of the grinding step. For some other biomass, the gain is more disputable. Finally, there are two existing markets: pre- and post-torrefaction. REVE unit has the benefit of being versatile and adapted for both applications, even in the same pellets production plant.

Whatever the technology is employed, the physical energy needed for torrefaction will be the same and can be estimated. For a biomass with initial moisture of 10–15%, the average minimum energy needed is 770 kJ/kg, or 215 kW h/ton [11]. The design of a REVE unit allows an energy efficiency of around 80–90% which means biomass torrefaction would require around 250 kW h/ton. Energy optimization could be done with the combustion of torrefaction gases, but the high amounts of water and CO_2 make it not easily feasible [11].

20.4 Conclusion

Combination of torrefaction and pelletization is a good process to produce solid biofuels with a high energy density and water-resistant, reducing the costs for bulk storage and transport. The choice between pre-torrefaction and post-torrefaction is not frozen and depends on feedstock and/or integration possibilities.

Both combinations have been validated into the REVE unit. Low torrefaction degree seems to be sufficient for increased grindability and hydrophobicity with 75 % of mass yield and 90 % of energy yield. Depending mainly on biomass moisture content, the minimum energy needed for torrefaction is approximately 215 kW h/ton, whatever the technology employed. Thanks to the high energy efficiency of the REVE, an industrial unit will consume approximately 250 kW h/ton. The simple and compact REVE unit could be easily integrated into a pellets production plant.

References

1. HOGAN, M. et al. (2010) *Biomass for heat and power—Opportunity and economics*, European Climate Foundation.
2. DOASSANS-CARRERE, N. et al. (2014) Comparative study of biomass fast pyrolysis and direct liquefaction for bio-oils production: product yield and characterizations. *Energy&Fuel* 28(8), 5103–5111.
3. TUMULURU, J.S. et al. (2011) A review on biomass torrefaction process and product properties for energy applications. *Ind. Biotechnol.* 7, 384–401.
4. REZA, M.T. et al. (2012) Pelletization of biochar from hydrothermally carbonized wood. *Environ. Prog. Sustain. Energy* 31, 225–234.
5. BERGMAN, P.C.A. et al. (2005) Torrefaction for biomass co-firing in existing coal-fired power stations. *Energy Res. Cent. Neth. Ecn Ecnc05013* 1–71.
6. SHANG, L. et al. (2012) Quality effects caused by torrefaction of pellets made from Scots pine. *Fuel Process. Technol.* 101, 23–28.
7. TUMULURU, J.S. et al. (2011) A review of biomass densification systems to develop uniform feedstock commodities for bioenergy application *Biofuels Bioprod. Biorefining* 5, 683–707.
8. PENG, J.H. et al. (2013) Torrefaction and densification of different species of softwood residues. *fuel.* 111, 411–421.
9. STELTE, W. et al. (2013) Pelletizing properties of torrefied wheat straw. *Biomass Bioenergy* 49, 214–221.
10. KOPPEJAN, J. et al. (2012) *Status overview of torrefaction technologies*, Enschede.
11. DOASSANS-CARRÈRE, N., MULLER, S., MITZKAT, M. (2014) REVE—a new industrial technology for biomass torrefaction: pilot studies. *Fuel Process. Technol.* 126, 155–162.